THE NEUROPSYCHOLOGY OF DEVELOPMENT:
A SYMPOSIUM

THE NEUROPSYCHOLOGY OF DEVELOPMENT:
A SYMPOSIUM

THE NEUROPSYCHOLOGY OF DEVELOPMENT

A SYMPOSIUM

EDITED BY

ROBERT L. ISAACSON

University of Michigan
Ann Arbor

JOHN WILEY AND SONS, INC.

New York · London · Sydney · Toronto

Library of Congress Catalog Card Number: 68-24796
SBN 471 42863 9
Printed in the United States of America

PREFACE

In 1967 the University of Michigan celebrated its one hundred and fiftieth birthday. In honor of its progressive development into a leading institution of higher education, the departments of the various colleges of the university sponsored special programs of interest to the university community. The Department of Psychology presented five such programs.

The Department's first symposium was called "The Scientific and Social Risks and Gains of Computer Technologies." Participants included H. A. Simon, J. C. R. Licklider, A. Rapoport, and D. G. Marquis, and its chairman was W. R. Reitman. The second symposium was entitled "Problems at the Frontiers of Social Psychology." Participants were T. M. Newcomb, L. Festinger, H. H. Kelly, W. J. McGuire, and H. C. Kelman. The planning chairman was R. B. Zajonc. The fourth program for the Psychology Department's sesquicentennial celebration was the John F. Shepard memorial lecture, "Leadership and Decision-Making in the Modern University," by R. W. Heyns, Chancellor of the University of California at Berkeley. Alfred C. Raphelson presented the high points of his study of "Psychology at Michigan: 1880-1950" at a Departmental Colloquium.

The papers collected in this book represent the contributions to the third symposium, "The Neuropsychology of Development." These papers are all original works undertaken especially for the symposium and contain many previously unpublished data and interpretations; the ideas expressed here are sometimes at odds with traditional views of the effects of physiological processes underlying development. We hope their challenge will motivate others to new work in the area.

In the first chapter Dr. Elliot Valenstein reviews material concerning the effects of steroid hormones administered at different times in the life of an organism. His interest in this area dates back to his graduate school training with Dr. William C. Young at the University of Kansas. In this paper, Dr. Valenstein emphasizes the importance of central nervous system factors in the regulation of pituitary and sexual hormone secretions, as well as the influence of hormones upon central nervous system structures. Especially interesting are the effects of prenatal administration of sexual hormones. He emphasizes that one of the permanent changes produced by steroid hormones may be a modification of the motivational influences of sensory information.

In conjunction with Leonard W. Schmaltz and Arthur J. Nonneman, I have been investigating the behavioral and anatomical effects of lesions made at various developmental stages of a "primitive" cortical area of the brain: the hippocampus. Our chapter in this book is our first written report of this work.

Our data suggest that lesions made shortly after birth are not without their effects upon behavior, even though they do not always produce the same results as adult lesions. In fact, early lesions sometimes produce behavioral changes which are in the opposite direction to those of adult lesions. Furthermore, the deficits following early lesions are entirely "task specific."

In the third chapter, Harry F. Harlow continues reports of a series of investigations on age-effects upon the deficits produced by frontal lobe lesions. New data are provided concerning the effects of frontal lobe lesions of different size on behavior when these lesions are made before the behavioral abilities of the animals become fully developed. The behavioral differences between animals with limited destruction of the lateral frontal surface and those with frontal lobectomies are interesting and unexpected. The finding that animals suffering frontal lobectomies within the first two months of life show impairment of learning set problems is remarkable, since it represents the only deficit of any type which has been found in Harlow's laboratory in animals suffering destruction of the frontal lobe areas at such an early age.

For a number of years I had known of Arthur Kling's interest in the amygdala and, more recently, in lesions of this area during infancy. In this chapter of this book, however, Dr. Kling and Dr. Thomas Tucker turn their attention to destruction of neocortical systems during infancy. One of the most significant of his results is the observation that the debilities on delayed-response and delayed-alternation problems are not equally relieved by the earliness of the brain damage, a result of special importance since it would indicate that the similar results found by Schmaltz, Nonneman and me, indicating the task-specificity of amelioration of deficits, may not be limited to the limbic system.

Eric Lenneberg's chapter represents a fitting conclusion to this book. His comments upon the many facets of the unusual pattern of human maturational history provides a framework against which the previous chapters may be profitably viewed. He presents data illustrating the effects of physiological variables upon behavior and the reverse effects. The concept of a critical period for these effects is stressed and illustrated by the results of several childhood diseases (e.g., German measles, kernicterus, and phenylketonuria) when they occur at different periods of development. The observations of how children "grow into" the symptoms resulting from early diseases points out the difficulties in early diagnosis of brain damage.

The work of the Department's committee which directed all of the Psychology Department's Sesquicentennial efforts should be recognized. The committee was composed of N. R. F. Maier, Judith Bardwick, Donald R. Brown, and John J. Brownfain; Helen Peak was its able chairman. Special appreciation is due Mrs. Nancy Bates and Mrs. Alexis Archibald, whose organizational efforts and administrative assistance with the organize of the present symposium and this book have been so helpful.

In his general comments of welcome before the Symposium began, Dr. W. J. McKeachie, Chairman of the Psychology Department, called attention to the Department's historical role in the area of physiological psychology. Among notable events he mentioned was the fire that was started by George H. Mead as he was shellacking a brain for laboratory demonstrations. It is the hope of all us associated with this book that the volume will help begin an "intellectual fire" directed toward more experimental attacks on the relationship between developing physiological and behavioral processes.

Robert L. Isaacson

Ann Arbor
January, 1968

CONTENTS

THE NEUROPSYCHOLOGY OF DEVELOPMENT:
A SYMPOSIUM

STEROID HORMONES AND THE

NEUROPSYCHOLOGY OF DEVELOPMENT

ELLIOT S. VALENSTEIN

Fels Research Institute
Antioch College

Although one could easily form a different impression from the number of texts and articles on the subject, *developmental psychology* represents an approach to problems rather than a distinct subject matter. Almost any psychological problem can be approached developmentally with at least two advantages resulting. First, by continually looking for the history behind a particular phenomenon and asking how it came about, we are more likely to raise the relevant questions and isolate the significant variables. Any attempt to explain "mouse killing" by cats, for example, without considering the development of that animal's response to moving objects in general, may grossly exaggerate the role of appetite satisfaction. The second advantage of an approach that stresses developmental considerations derives from an appreciation of the fact that stimuli and responses, or more generally variables and their consequences for an organism, cannot be considered independently. Organisms are constantly changing with each interaction with the environment, and as a result we cannot study the effects of any environmental condition without considering the nature of these changes Thomas Wolfe must have had a global appreciation of this concept when he chose the title for his novel *You Can't Go Home Again*.

During the initial exploration of a problem, *developmental* psychology consists of a descriptive account of the order or sequence of events during the life of an organism, or it may consist of a listing of the consequences of a specific experimental manipulation at different ages. The participants at this symposium are well aware that determining the effects of brain lesions at different ages is only the first step. What is needed is some understanding of the behavioral and morphological processes which account for the different deficits produced by early and late brain lesions. Often this can be achieved only through a series of relevant changes in the testing situation designed to determine what is and

The support of Research Grant M-4529 and Career Scientist Award MH-4947 from the National Institutes of Health and Research Grant NsG-437 from NASA in the preparation of this manuscript and in conducting some of the studies by the author and his collaborators cited in the text is gratefully acknowledged.

1

what is not learned. The work of Gollin (27) with normal children of different ages provides an example of this kind of behavioral analysis.

Gollin tested children 3½ to 4 and 4½ to 5 years of age in a two-pattern discrimination problem and a reversal problem in which the positive and negative patterns were interchanged. If the children were given additional practice on the initial task after criterion had been achieved, it was found that the younger but not the older children took significantly longer to learn the reversal problem. For the younger children the amount of interference in learning the reversal problem was directly related to the amount of overtraining. Previously it had been found that even without overtraining, the younger but not the older children exhibited negative transfer in the first block of trials on the reversal problem (29). These results and others suggested to Gollin that the younger children were dominated by the stimulus properties of the task, whereas the older children, who shifted very rapidly regardless of the amount of overtraining, responded to the conceptual properties of the task (28). The fact that these constructs may be difficult to deal with physiologically may reflect more on the present state of physiology than on any weakness of the formulation. In any case, there was sufficient evidence that the children of different ages did not learn the same information when faced with the same task. A full appreciation of the implications of such an approach is necessary for any understanding of the significance of the variation in results produced by brain lesions at different ages as well as those cases in which symptoms remain hidden until some later stage of development.

My contribution to this symposium is developmental in two senses. In part the plan is to present an abbreviated chronology of our changing views of the action of steroid hormones on the nervous system and behavior because of the belief that the perspective gained would produce a more balanced view of our present conceptual framework. In addition, the brief introductory comments on the effects of variables at different times in the life history of an organism will be expanded by considering the action of steroid hormones in this context. I wish that more in the direction of processes and mechanisms could be offered at this time, but the present state of the art does not make this possible.

To begin, let us go back to the period bracketing World War II. It was well established by that time that the sexual behavior of at least lower mammals was completely dependent on the presence of gonadal hormones. The gonadectomized animal exhibited a clear drop in sexual interest and capacity, although it was usually stated that the sexual behavior of the female, but not that of the male, completely disappeared following gonadectomy. With respect to such animals as the rat and guinea pig these findings are too well known to need documentation. The interested reader may refer to the excellent review by Young (101).

However, many details about the quantitative relationship between hormones and sexual vigor were not known. In the laboratory of Dr. William C. Young during the early 1950s a quantitative method of scoring the sexual vigor of the male guinea pig was developed (104). Similar techniques were developed for other species in other laboratories. In general, the scoring procedure involved recording discrete elements of sexual behavior, and by giving an appropriate weighting to latencies and the maturity of the behavior a composite score of sexual vigor could be assigned. It was found that individual differences were quite reliable over repeated tests (105) and the question of the basis of these differences quite naturally arose.

Grunt and Young (40, 41) studied the possibility that the quantity of androgen was responsible for these differences. In Figure 1 the results of a part of this work can be seen. During the preoperative period, prior to gonadectomy, clear differences were obtained between so-called high-, medium-, and low-drive animals. Following castration, sexual behavior dropped to a minimum and the differences between the animals disappeared. With replacement hormonal therapy (testosterone propionate) the animals separated out in accordance with their preoperative levels. Low hormonal doses were not adequate to restore the animals to preoperative levels, whereas injections four to eight times the adequate dose did not eliminate the differences between the high-, medium-, and low-drive animals. Some qualitative and quantitative differences in the sexual behavior of the male rat receiving very high doses of androgen had previously been reported by Beach and Holz-Tucker (9), but the data were not of a nature to alter these conclusions. For example, although high doses shortened latencies in sexual responding, the number of incomplete copulations increased.

Grunt and Young concluded that the initial differences in sexual vigor could not be accounted for on the basis of differences in hormonal production. Subsequently, Dr. Robert Goy (33, 34), working in the same laboratory, quantified the sexual behavior of the female guinea pig. Primarily this consisted of a measure of the latency and duration of the lordosis responses. Reliable individual differences were found with the female which were not dependent on quantities of hormone (in this case estrogen and progesterone), and the authors concluded that "In general, therefore, the principle established for the male, that supraliminal quantities of gonadal hormone do not alter the characteristic pattern of the behavior, can be extended to the female" (34, p. 352). Earlier, Young and his colleagues had failed to find any correlation between the number of rupturing follicles and the duration of the lordosis response (102, 106).

Although the amount of hormone available to an adult animal did not seem to be a major determinant of the sexual vigor, it remained possible that quantity of hormone was a significant variable during development. Figure 2 illustrates results from a study addressed to this problem by Riss, Valenstein, Sinks and

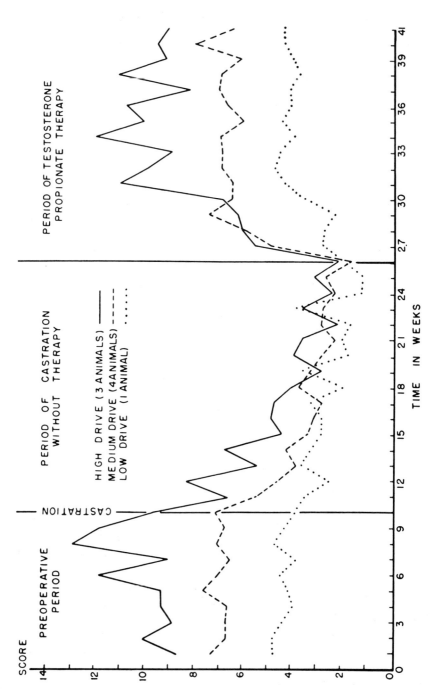

Figure 1. The effect of castration and therapy with 50γ testosterone propionate daily on the sex drive of high-, medium-, and low driven male guinea pigs. [After Grunt and Young (40).]

Young (84). Males from two highly inbred strains and a sexually vigorous heterogeneous group were used. The sexual performance of intact males from these groups differed as seen in the figure. Representatives from the three groups were castrated at birth and given 500 μg of testosterone propionate per 100 gm of body weight daily. This is a large dose. The results are clear. The behavior of the castrates given replacement hormone, when tested as adults, resembled that of the genetic group from which they originated.

Still another experiment bears on the topic under discussion. Valenstein, Riss, and Young (96, 97) had found that the adult male guinea pig required some contact with other animals for the full sexual pattern to be expressed. Males not provided with adequate contact did not score as high in the sexual behavior tests, not because they were not sexually oriented and aroused, but because they were not able to mount and clasp the female in a fashion that permitted the culmination of the sexual act. Some controversy surrounds the exact nature of the deficit and whether male rats also require some experience (5, 8, 107), but for the point under discussion it is only necessary to demonstrate a difference in sexual behavior pattern in one species between socially reared and isolated animals. On the left of Figure 3 the difference in the sexual behavior score between the socially reared and isolated animals is displayed. Half of each of these groups was castrated and given several tests a number of weeks later to establish that the animals had shown the customarily observed decrease in sexual behavior following castration and that the initially observed differences had disappeared. Following hormonal replacement therapy the sexual behavior of the castrates returned to that of the preoperative level, illustrating that the animals were performing quantitatively and qualitatively at the same level as the intact controls. Even with additional hormonal administration the isolated animals did not exhibit the complete sexual pattern.

The conclusion drawn from these experiments and others that have been reviewed by Young (101) was: *Steroid hormones are necessary to activate the neural substrate underlying sexual behavior, but the character of the substrate determines the quantitative and qualitative aspects of the behavior expressed.* When talking informally, those of us working in Dr. Young's laboratory at this time commonly used an analogy from the photographic laboratory to explain our conception of the interaction between steroid hormones and behavior. It was said that the hormone is comparable to a developer in that it was necessary to bring out the picture, but in spite of the fact that it was possible to over-develop or underdevelop, the picture could not be radically changed by the amount of developer or the time that it was used.

Although I have emphasized that it was the neural substrate mediating behavior on which the hormones acted, the relevance of hormonal influence on peripheral structures was also being indicated by research during this same period. For example, hormones acting on the crop of the ring dove (61, 62) or

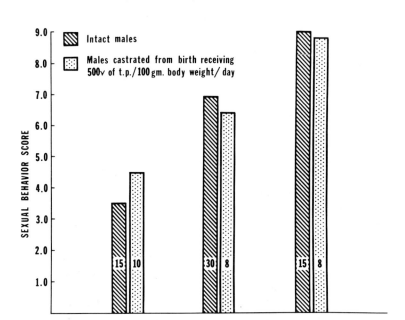

Figure 2. Sexual behavior scores of intact males with three different genetic backgrounds and males castrated from birth receiving testosterone propionate replacement therapy. [Adapted from data obtained by Riss, Valenstein, Sinks, and Young (84).]

the glans penis of the rat (10) might set up "drive stimuli" which could direct the development of behavior of the organism interacting with its environment. Presumably, the central nervous system would be affected by afferent feedback, but this would represent an indirect effect of hormones on the central nervous system. That such peripheral factors as an engorged crop may be important for the development of parental feeding in the ring dove was suggested by Lehrman's demonstration that feeding of the squab is inhibited by local anaesthesia of the crop (62).

A number of lines of investigation in recent years have provided convincing evidence that the steroid hormones exert a direct influence, and probably their major influence, on central nervous system structures. Although for many years the relationship of the gonads to the anterior pituitary was believed to be that pictured on the left of Figure 4, there was really evidence available for some time which should have subjected this view to serious questioning. The scheme depicted and the one taught to me during my own training (so it is really not ancient history) was that the trophic hormones of the pituitary and the hormone of the target organ had a direct feedback relationship. At that time my teacher,

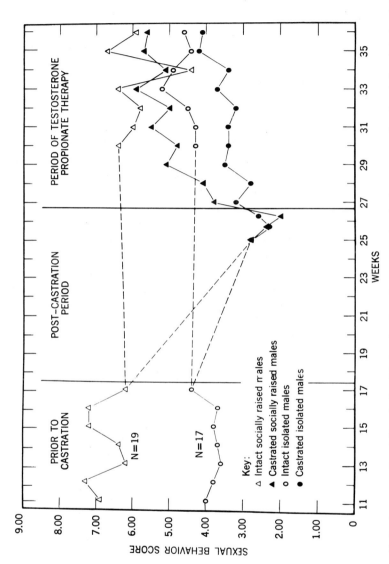

Figure 3. Sexual behavior scores of socially raised and isolated males prior to castration, in the postcastrational period, and during the period of testosterone propionate therapy. [After Valenstein and Young (97).]

7

Will Young, was still reflecting the thinking of his own teacher, Carl Moore (76). Pituitary gonadotrophic hormone *stimulated* the release of gonadal hormones, and circulating gonadal hormones were said to directly inhibit the further release of trophic hormone. The fact that castration produced abnormally enlarged basophilic cells, so-called castration cells, in the anterior pituitary seemed to demonstrate the effects of removal of a negative feedback loop.

The description by anatomists that there were only vascular, but not neural, connections between the nervous system and the anterior pituitary supported the view that hormonal levels in the blood exerted their influence directly on the anterior pituitary (37). This view was far from adequate. It was not at all easy to explain how this servomechanism permitted the departure from what should have been relatively stable hormonal levels when the environment in the form of cold, heat, light, dark, and so forth, placed some special demands on the organism. Furthermore, the attempts to explain the various cyclicites in hormonal levels in the female, often very imaginative, were quite strained and supported by little fact. Still further, the observations that motivated such songs as "They're Laying Eggs Now Just Like They Used To Ever Since That Rooster Came Walking By" could not all be dismissed as anecdotal. In this context, Marshall (67) had suggested over 30 years ago that the function of sexual display in many birds was to insure synchronization of male and female reproductive processes, and Matthews (70) shortly afterward had demonstrated the effectiveness of a visual stimulus by finding that even the introduction of a mirror into

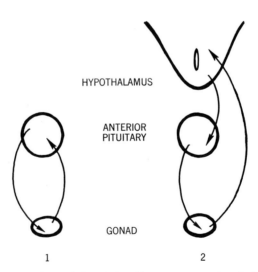

HYPOTHALAMUS

ANTERIOR
PITUITARY

GONAD

1   2

Figure 4. Early and later views of the relationship between anterior pituitary gonadotrophic secretions and ovarian hormones.

the cage of isolated pigeons increased egg laying. The large accumulation of evidence indicating that anterior pituitary functioning was influenced by stimuli which required the analytical power of the nervous system for processing has been reviewed elsewhere (47).

Later work demonstrated that the major influence of the gonadal hormones must be exerted directly on the brain, which in turn influenced the activity of the anterior pituitary by its own humoral control, now called "releasing factors" (71), as shown on the right side of Figure 4. Work with central nervous system drugs by Everett, Sawyer, and Markee (22, 88) demonstrated that compounds that did not effect glandular discharge directly could block the release of anterior pituitary trophic hormones by their action on the nervous system. Earlier, Harris (46) and Haterius and Derbyshire (54) had demonstrated that ovulation could be elicited by electrical stimulation of the hypothalamus. These data clearly pointed to the way that endocrine cycles could be synchronized with environmental rhythms, and by extrapolation the influence could be extended to almost any "business" of the nervous system.

Figure 4 should be expanded to denote the manifold influences mediated by the central nervous system which may modulate the hypothalamo-hypophyseal relationship. The reader who has a predilection for pictorial schema illustrating the complex interactions between the gonads, pituitary, hypothalamus, environment, and behavior may refer to the book by the Scharrers (89). Furthermore, in emphasizing the importance of the hypothalamus in the feedback circuit it should not be assumed that a coexisting direct feedback route from gonad to anterior pituitary has been completely ruled out by the evidence. A comprehensive and balanced review of the evidence in this field may be found in recent reviews by Donovan (20) and Everett (21). The transplanting of ovarian tissue directly into the hypothalamus and noting that gonadotrophic hormones were inhibited while transplanting the same tissue into the pituitary did not inhibit trophic hormone release provided further support for the conclusion that the feedback circuit from gonad to pituitary probably involved the intermediary action of the hypothalamus (25). Later, Lisk (66) demonstrated that small amounts of estrogen in the posterior hypothalamus sharply reduced the amount of FSH and LH secreted by the anterior pituitary, and Michael (72) reported that systemic injections of labeled estrogens were found to be taken up selectively by hypothalamic sites.

More recently, evidence was provided that small amounts of hormone may elicit behavior by direct application to central nervous system structures. Barfield (1), for example, demonstrated that minute amounts of crystalline testosterone propionate administered through cannulae directly into the brains of capons elicited copulatory behavior. The injected capons regularly mounted hens and exhibited treading and tail depressing responses in the normal fashion. Particularly significant was the fact that comb growth, which provides a very sensitive measure of androgen in the systemic circulation, was not correlated with the occurrence of copulatory behavior. Apparently there are androgen-sensitive regions in the chicken brain capable of eliciting male sexual behavior when the appropriate stimulus is present. Similarly, it was demonstrated that estrous behavior in the ovariectomized cat could be activated by estrogens administered directly into the hypothalamus in quantities too small to produce peripheral anatomical change, that is, changes in the genitalia (52, 53, 73). Earlier, Fisher (23) had reported the elicitation of maternal and sexual behavior with chemical stimulation, but little information was available on the state of peripheral structures. These studies indicated in a convincing fashion that the gonadal hormone influence, both on pituitary trophic hormone activity and on sexual behavior, was primarily indirect with the nervous system serving as an active "middleman."

The preceding may be viewed as background for two other scientific streams, which had independent sources but were converging. A classic paper by Pfeiffer (80) in the thirties had shown that ovaries transplanted into male rats exhibited cyclical functioning and corpora lutea formation only if the males were castrated at birth. The presence of androgen for any appreciable time after birth, even if then removed, would cause the male to lose this ability to control cyclical functioning. Pfeiffer guessed incorrectly that the pituitary of both the male and female was undifferentiated at birth but androgen differentiated the pituitary of the male, and as a result it lost the capacity to secrete gonadotrophins in a cyclical fashion. Pfeiffer was correct, however, in speculating that androgen played a differentiating role; the mistake was in locating the site of the differentiation in the pituitary.

A number of later pituitary transplant studies from males into females revealed that the pituitary remained plastic and pluripotential. Harris and Jacobsohn (50), for example, reported that the pituitaries from male rats transplanted beneath the median eminence of hypophysectomized females were capable of supporting normal estrus cycles, mating, pregnancy, and lactation. This work was later confirmed by Martinez and Bittner (68) in mice. The plasticity of the pituitary is also illustrated by the demonstration of Kallas (57, 58, 59) that the pituitary from an immature animal could support adult gonadal functioning. Kallas worked with a parabiotic preparation in which the increased gonadotrophic hormone production from an immature, castrate animal was shown to accelerate the occurrence and later support of mature gonadal functioning in a chronologically immature animal. Harris and Campbell (49) and Donovan (20) have presented excellent reviews of this work.

In the study by Everett, Sawyer, and Markee (22) referred to earlier, the authors concluded from their work with drugs affecting the central nervous system that there was a hypothalamic "center" controlling cyclical activity. With their own results in mind and referring to the study of Pfeiffer, they prophetically wrote:

"It appears that in rats during infancy the action of androgen conditions differentiation of the hypothalamic center as an intrinsically acyclic mechanism. However, in an intact genetic female or in a male castrated in infancy and implanted with ovaries, the center differentiates as an intrinsically cyclic mechanism. . . . It now seems probable that this sex difference actually resides in the hypothalamus" (22, p. 248).

Although it is always difficult to establish priority, and there were several earlier statements by others which may be viewed as containing a germinal element of the same thought,* the article by Everett *et al.* (22) seems to contain the first clear statement of this position. *The implication of this view is that with respect to the hypothalamo-anterior pituitary control of gonadotrophic hormone release, steroid hormones play an organizational role but this role is restricted to an early period of development.*

What about behavior? The studies cited had emphasized *the activational role* of the hormone and *the significance of the neural substrate,* whether shaped by genetic or experiential factors, as a determinant of the quantity and quality of the elicited behavior. The significance of the substrate emerged also from a review of the literature concerning the ability of androgen to elicit male behavior in females and estrogen-progesterone to elicit female behavior in males (101). Here, too, the morphological base upon which the hormone acts seems to be the most significant variable. In the case of either the male or female the elicitation of the opposite behavior was usually weak and fragmentary and the amount of hormone required was well above that necessary with the opposite sex and usually well above a physiologically meaningful dose. Therefore, up to this point in the chronology there was evidence for only an activational role of gonadal hormone with respect to behavior.

Probably in part because Vera Dantchakoff was working with the guinea pig, which was at the time the favorite experimental animal of the group associated with Will Young, her results were particularly meaningful to these workers. Dantchakoff (16, 17, 18, 19) had reported that female guinea pigs born to mothers given androgen injections (directly into the amniotic cavity) during the gestation period were masculinized anatomically. These genetic females possessed a well developed epididymis, prostate and Cowper's gland, and a penis in addition to an ovary and normal female genital tract. A bibliography of similar findings with the mouse, rat, hamster, hedgehog, mole, rabbit, cattle, goat, and monkey is summarized in tabular form by Grumbach and Ducharme (39), who also present

---

*Earlier, Holweg and Junkmann (55) had suggested a "sex centrum" in the hypothalamus after noting that pituitary gland transplanted at a site separated from the hypothalamus did not develop castration cells after gonadectomy. Many workers had speculated earlier on the possibility of neural influence on endocrine function. The novelty of the suggestion by Everett, Sawyer and Markee (22) was in raising the possibility that the hypothalamus of the genetic male and female may be different.

case histories of androgen-induced female pseudohermaphrodism in humans. Preliminary results suggested that these masculinized females when given androgen performed sexually as males. There was also available for some time information that the freemartin (a female calf usually sterile that was twinborn with a male) often exhibited male behavior. There was, of course, speculation that the female calf was masculinized *in utero* by the circulating androgen released by her twin.

Phoenix, Goy, Gerall, and Young (81) undertook the first systematic experimental study of the sexual behavior of masculinized females (hermaphrodites). In this experiment the female offspring were masculinized by injecting (subcutaneously) the pregnant sows with testosterone propionate. Table 1 presents some of the results of this study and illustrates that in terms of the amount of male behavior (mounting) exhibited during tests with females and the amount of androgen necessary to elicit this behavior, the hermaphroditic female was closer to the male than the untreated female. It is important to emphasize also that in the same study it was found that only a very low level of female behavior could be elicited by estrogen in combination with progesterone. The hermaphrodite therefore not only exhibited significantly more male behavior but also significantly less female behavior than the normal female. The behavior of the male offspring born to the androgen treated mother was indistinguishable from that of untreated male guinea pigs. Other experiments from this laboratory determined the critical prenatal period when androgen was most effective in differentiating the nervous system of the female guinea pig into the male pattern (32).

Since rats have short gestation periods, it was possible to do a number of experiments with this species that would be more difficult with a relatively long-gestation animal like the guinea pig. The nervous system of the male rat is apparently less differentiated at birth than that of the guinea pig, and castration of the newly born rat produces an adult animal that exhibits significantly more female behavior and significantly less masculine behavior in response to appropriate hormone administration than the male castrated at a later time (35). In the case of the female rat, again because of the short gestation period, it was possible for Harris and Levine (51) to masculinize the female by a postnatal injection of androgen during the first 5 days after birth. More striking results

TABLE 1. MASCULINE BEHAVIOR IN GONADECTOMIZED ADULT ANIMALS INJECTED WITH TESTOSTERONE PROPIONATE

Group	Mean sexual behavior score	Mean mounts per test	Median number of tests to the first display of mounting	Median mg. of t.p. prior to the display of mounting
Spayed untreated females	2.1	5.8	7.0	30.0
Spayed hermaphrodites	3.6	15.4	3.0	10.0
Males castrated prepuberally	5.0	20.5	1.5	3.8

After Phoenix, Goy, Gerall, and Young, (81)

have been obtained by Gerall and Ward (26), who have administered androgen prenatally to female rats, which exhibited (when tested as adults) significantly more male behavior and less female behavior in response to hormonal treatment than normal females.

It is important to appreciate that in both the guinea pig and the rat, androgen exerts a double action on behavior. When androgen is present at critical times the female later exhibits significantly more male behavior and less female behavior. When androgen is absent at critical times the male later exhibits significantly more female behavior and less male behavior. Estrogen has a different effect. Early administration of estrogen to the female may produce a sterile animal (3) which exhibits less female behavior (99) but not necessarily more male behavior. Similarly, early administration of estrogen to the male interferes with the normal development of male behavior but does not necessarily produce more female behavior. The evidence is not uniformly in agreement with the preceding summary, as Levine and Mullins (63) have reported that estrogen administered to neonatal female rats may increase the amount of male behavior displayed during adulthood; administration of the same dose of estrogen to neonatal male rats did not interfere with display of male behavior if allowance was made for the lack of development of the accessory sexual apparatus. In spite of these conflicting results, the evidence in general supports the conclusion that estrogen administered at the wrong time has a disruptive or deleterious effect rather than an organizational role. The major conclusions stemming from this work are the following:

The neural substrate mediating sexual behavior goes through a relatively undifferentiated stage. During this period androgen is effective in differentiating the nervous system in a male direction and the information now available indicates that this trend is irreversible. In the absence of androgen, development proceeds in a female direction. Estrogens do not appear to play an active role in imposing a female organization on the developing nervous system. Some additional support for this position stems from experiments demonstrating that even in short-gestation animals like the rat, females ovariectomized at birth respond to estrogen and progesterone injections when adult by exhibiting normal female sexual behavior (quantitatively and qualitatively). The results seem to parallel the embryological data on the differentiation of morphological structures in that androgen is the active differentiator. Female development depends more on the absence of androgen than the presence of estrogen.* The reader is referred to an excellent review of the literature on morphological differentiation by Burns (14). In 1945, Beach (6) had reported a case of a female rat with a congenital absence of any

---

*Here, too, there exists some conflicting evidence that suggests that dose levels may be a complicating variable. Burns (13) and Moore (75), for example, have noted that large doses of testosterone propionate may stimulate markedly the Müllerian ducts (the anlage of the female genital tract) of the female opossum.

ovarian tissue. The reproductive anatomy in this animal was clearly female, but remained at an infantile, postnatal level. Following injections of estradiol benzoate and progesterone this animal displayed the complete female mating pattern when placed with a sexually vigorous male. On the other hand, the active role of androgen has been demonstrated recently by Neumann and Elger (79), who have treated pregnant rats with the anti-androgen cyproterone acetate. In the "feminized" male offsprings that were castrated, nipples were developed and it was found that the mammary glands reacted to estrogen-progesterone treatment in a manner similar to female animals. In castrated males which had not received the anti-androgen prenatally, nipples were not developed and the mammary glands did not exhibit any proliferation in response to hormonal treatment. This report is consistent with earlier work by the same authors (77,78) demonstrating that prenatal administration of anti-androgens prevented the differentiation of male genitalia and produced genetic male rats with a female sexual orientation and cyclic gonadotrophic release.

It became evident as a result of the work with early administration of steroid hormones that these hormones played more than an *activational* role with respect to behavior. When administered early, androgen could influence the pattern of behavior subsequently expressed and therefore was playing an *organizational* role. The evidence and arguments were becoming increasingly convincing that the development of the nervous system could be qualitatively changed by the action of androgen at early critical periods. This appeared to be true not only for the nervous system's control over anterior pituitary gonadotrophic hormone functioning, but also with respect to the organization of the neural substrate underlying sexual behavior.

More recent information suggests that the preceding conclusions may be more far-reaching and extend beyond the limits of what is normally considered sexual behavior. In Figure 5 the normal amount of daily activity of a male and female rat are shown in the top two graphs (48). As is well known, the activity of the female is much greater than that of the male, and, in addition, the female's activity peaks about the time of estrus. The bottom two graphs depict the activity of males castrated on the day of birth and given ovarian and vaginal transplants at 63 days of age (denoted by the T on the graphs). The vaginal transplants were placed under the abdominal skin according to the method of Yazaki (100). After the transplants, activity increased dramatically and tended to peak at the time of estrus, which could be detected from the smear picture obtained from the vaginal transplants. Males given ovarian transplants but not castrated early did not respond in this way. The interpretation may be complicated by Kennedy's report (60) that administration of androgen to female rats during either the last 5 days of gestation or the first 6 days after birth does not reduce running activity.

Perhaps partially related to the activity picture, although believed to be primarily a measure of emotionality and autonomic reactions, is the behavior seen

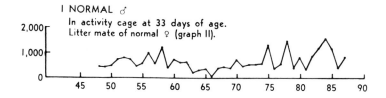

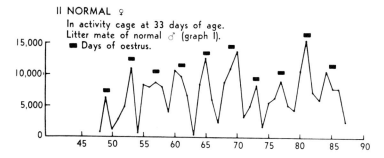

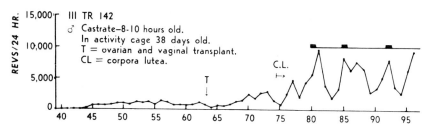

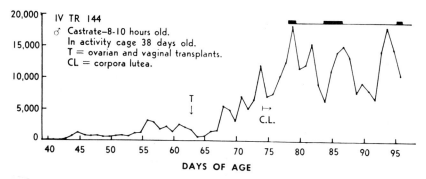

Figure 5. Activity of normal male and female rats and neonatal castrate males with ovarian transplants. [After Harris (48).]

in the open-field test (12). This test is simply a measure of the amount of exploration (or, conversely, "freezing") and the amount of defecation displayed in a highly illuminated arena under standardized conditions. Figure 6 presents some data collected by Kerry Drach while he was doing a senior undergraduate study in our laboratory. Drs. Verne Cox and Jan Kakolewski also collaborated on this work and other experiments from our laboratory to be presented subsequently. Figure 6 illustrates the open-field behavior (grids crossed and amount of defecation) of male and female rats, half of which received an injection of 500 to

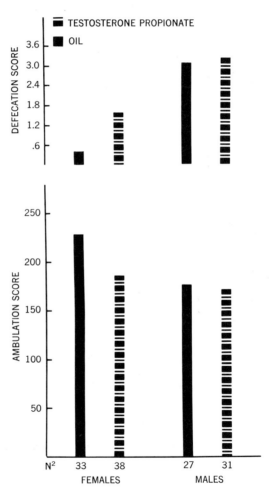

Figure 6. Open-field behavior of normal male and female rats and animals receiving an injection of testosterone propionate on the third or fourth day of life. (Unpublished data from Drach, Cox, Kakolewski, and Valenstein.)

1000 µg of testosterone propionate on the third or fourth day after birth. The other half was given oil injections. It can be seen that the oil-injected animals exhibited the usual sex difference found in this test. The untreated females ambulated more and defecated less. The females which received the androgen injection, however, behaved more like the males, ambulating less and defecating more than normal females. To the extent that this test reflects emotionality in general the results could be viewed as most dramatic. Similar results were obtained previously by Gray, Levine, and Broadhurst (36) in the rat and by Swanson (91, 92) in the hamster.

There are also weight and growth differences between males and females of most species. These are very striking in the case of the rat. Figure 7 shows the weight record of the animals whose open-field behavior was presented in Figure 6. Since this experiment was run in two platoons and the growth records were different, the data have been charted separately. In both cases the males were significantly heavier than the females and there were no differences between the oil- and androgen-treated males. However, the female animals that received the testosterone propionate injection were significantly heavier than the female control animals, although still not equal to the males in weight. Barraclough (2) ruled out the possibility that the protein anabolic effect of androgen was the sole cause of the weight change as female animals injected at 10 to 20 days of age grew normally. Harris (48) has suggested that early androgen treatment may produce a sexual differentiation in neural control over growth hormone secretion. Whether or not the extra weight the androgen treated females "could throw around" had any behavioral consequences is a matter of conjecture at this point.

We were searching for other behavioral tests in which a sex difference might be demonstrated. In another context we had been studying the sex differences in hyperphagia and body weight change following ventromedial hypothalamic damage (95). With equivalent damage to the ventromedial hypothalamic area the females display significantly greater increases in food consumption and body weight changes than do the males. Figure 8 shows the results of such an experiment with animals matched for age. Under these conditions the males are much heavier. It can be seen that only the operated females, but not the operated males, exhibit a dramatic increase in body weight gain and food consumption over their anaesthetized controls. Figure 9 shows the same results with groups of male and female animals matched to the weight of the opposite sex in Fig. 8, while Table 2 presents these results quantitatively.

Elsewhere we have reviewed and I think effectively eliminated the argument that the males are normally consuming food and increasing their weight at rates close to maximal levels and therefore are not able to display any further significant increases in these measures. For example, Figure 10 illustrates the results of an experiment by Steinbaum and Miller (90), who have demonstrated that mature male rats stimulated for only 2 hours daily in the lateral hypothalamus

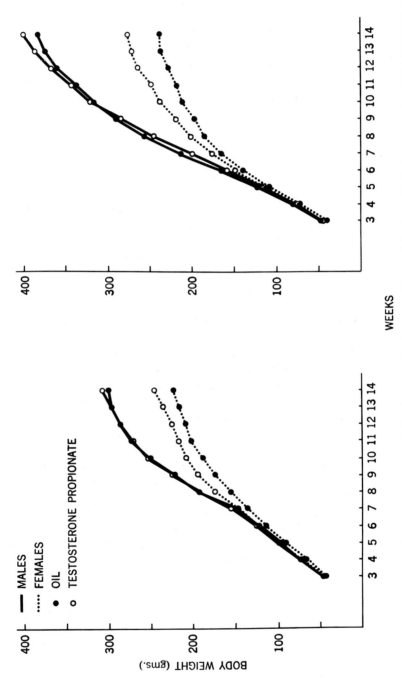

Figure 7. Body weight curves of normal male and female rats and animals receiving an injection of testosterone propionate on the third or fourth day of life. (Unpublished data from Drach, Cox, Kakolewski, and Valenstein.)

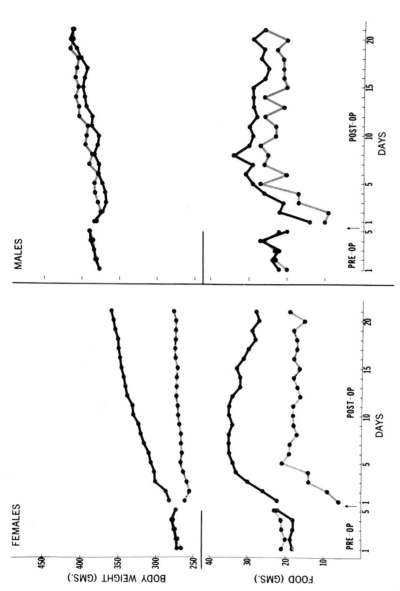

Figure 8. A comparison of the changes in body weight and food consumption of VMH lesioned males and females matched for age. Results of lesioned animals are depicted with solid lines, control animals with broken lines. [After Valenstein, Cox, and Kakolewski (95).]

19

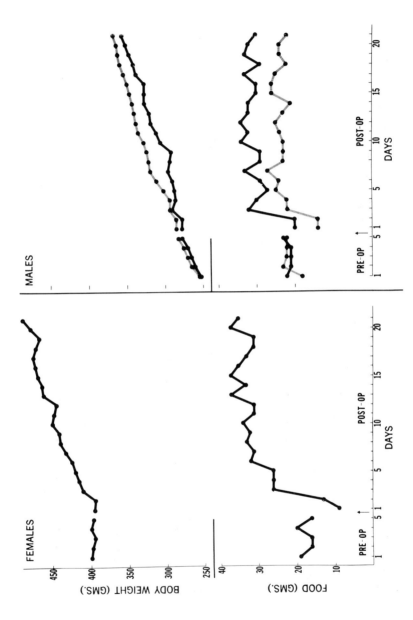

Figure 9. A comparison of the changes in body weight and food consumption of VMH lesioned males and females matched to the body weights of the opposite sex in Fig. 8. [After Valenstein, Cox, and Kakolewski (95).]

TABLE 2. COMPARISON OF THE EFFECT OF VENTROMEDIAL HYPOTHALAMIC LESIONS ON FOOD CONSUMPTION AND BODY WEIGHT IN MALES AND FEMALES

	Light Females N = 8	Heavy Females N = 9	Light Males N = 7	Heavy Males N = 12
Av. Pre-op Body Weight (gms.)	276	396	278	390
Av. Pre-op Daily Food Consumption (gms.)	19	17	22	23
Av. Post-op Daily Food Consumption (gms.)	31	30	30	27
Av. Pre-op Daily Weight Gain (gms.)	.5	-.9**	6.4	3.1
Av. Post-op Daily Weight Gain (gms.)	4.0	4.7	3.8	1.4
% Increase in Av. Food Consumption	63%	76%	36%	17%
% Increase in Body Weight*	26%	17%	-13%	-10%

* The formula $\dfrac{\text{FINAL WT.} - \text{PROJECTED WT.}}{\text{PROJECTED WT.}} \times 100$ was used. The projected weight was based on the average pre-operative weight gain.

** The projected weight was calculated by the more conservative assumption of no weight gain rather than a loss.

21

consume over 100% more food per day than nonstimulated control animals. In addition, these workers have demonstrated that during the period of lateral hypothalamic stimulation for only a 2-hour period each day, male rats gained an average of 7.5 gm daily compared to 2.2 gm for nonstimulated controls. The effects persisted for over 20 days and the authors concluded that there was no indication that a maximum had been reached. Furthermore, in our laboratory as well as at a number of other institutes, the food consumption of male rats has been doubled by daily injections of a long-acting insulin. There seems to be little doubt that the male rat is capable of large increases in food consumption and rate of body weight gain, but these are not seen following ventromedial hypothalamic lesions as frequently as in the case of females.

In collaboration with Kerry Drach and Drs. Cox and Kakolewski, we attempted to determine if early androgen administration would modify the sex

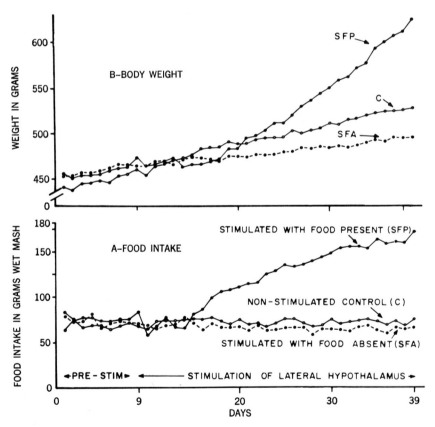

Figure 10. Effect of hypothalamic stimulation on A: food intake; B: body weight. Group means shown. [After Steinbaum and Miller (90).]

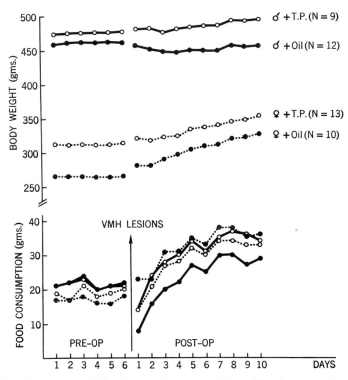

Figure 11. Comparison of the changes in body weight and food consumption of VMH lesioned male and female oil-injected rats and animals receiving an injection of testosterone propionate on the third or fourth day of life. (Unpublished data from Drach, Cox, Kakolewski, and Valenstein.)

differences generally observed following ventromedial hypothalamic lesions. As indicated, male and female rats were given 500 to 1000 μg of testosterone propionate on the third day of life. Approximately half of the animals served as controls and were given only oil injections. Figure 11 illustrates the effectiveness of the ventromedial hypothalamic lesions in producing increases in rate of weight gain and food consumption in normal and testosterone propionate injected males and females. The data are charted for 6 days prior to and 10 days after the VMH lesions. Table 3 presents the results for this same period quantitatively. The results should be viewed as suggestive rather than conclusive, and we plan to replicate the experiment. In general, the testosterone-treated males were not very different from the normal males. As is characteristic of the males, there were not striking changes following the VMH lesions. There was a slight tendency for the testosterone-treated animals to show a little more weight gain and increase in food consumption than the control males. It is possible that this difference is reflecting an inhibition of masculine differentiation caused by exogenous androgen. A little support for this interpretation can be gained from the reports that androgen may have an inhibitory effect on the developing testes, and indeed in the present experiment the testes of the males treated with androgen were significantly lighter than those of the control animals in spite of a heavier body weight.

The "masculinized" females, on the other hand, did not exhibit as great an increase in rate of weight gain or food consumption following the VMH lesions as did the normal females. In this respect, the "masculinized" females were more like the normal males. This can be seen by comparing the slopes of the lines illustrating body weight following the VMH lesions and from the fact that whereas the "masculinized" females were eating slightly more than the normal females prior to the lesions, this order was reversed after the lesions (Figure 11). This same trend can be seen by comparing the percentage of increase in rate of body weight gain and food consumption presented in the last two rows of Table 3.

The preceding groups of animals were tested with two other procedures that normally produce a difference between male and female rats. Valenstein, Cox, and Kakolewski (94) had reported a sex difference in taste preference between saccharin and glucose solutions. In addition, we determined if the commonly observed sex difference in reaction to pentobarbital could be influenced by prenatal administration of testosterone propionate. For this purpose we employed a "sleep test" similar to that used by Levitt and Webb (64, 65). Females normally succumb to the pentobarbital sooner and sleep significantly longer than males. Although the results of the taste experiment showed a trend for the "masculinized" females to exhibit preferences similar to the male, the data were only suggestive and not statistically significant. The females receiving neonatal testosterone propionate, however, responded as normal females in the pentobarbital sleep test.

Recently, Conner, Levine, Wertheim, and Cummer (15) reported that neonatally castrated male rats responded similarly to females in a shock-induced fighting situation. Normally, in intact animals, males exhibit significantly more fighting in response to shock than do females. Furthermore, the neonatal castrates were similar to the females in that they did not increase the amount of fighting displayed in response to injected androgen. The authors have concluded tentatively that in neonatal castrates the neural substrates on which androgen acts in modulating "aggressive behavior" are organized in a female pattern.

The work with monkeys by Goy, Phoenix, and Young (31, 82, 103) represents the greatest extension of the behavioral implications of the organizational role of steroid hormones. Following a procedure developed earlier by Wells and van Wagenen (98), genetic female rhesus monkeys were masculinized by injecting testosterone propionate daily into the pregnant mothers from the thirty-ninth to approximately the seventy-fifth day of gestation. The genitalia of the pseudo-hermaphroditic females that were produced were characterized by a well formed scrotum and a small, incompletely formed penis. The external vaginal orifice was

TABLE 3. COMPARISON OF THE EFFECT OF VENTROMEDIAL LESIONS ON FOOD CONSUMPTION AND BODY WEIGHT IN TESTOSTERONE PRCPIONATE OR OIL INJECTED MALES AND FEMALES

	Females + Oil N = 10	Females + T.P. N = 13	Males + Oil N = 12	Males + T.P. N = 9
Av. Pre-op Body Weight (gms.)	266	315	462	478
Av. Pre-op Daily Food Consumption (gms.)	17	19	22	22
Av. Post-op Daily Food Consumption (gms.)	32	29	23	30
Av. Pre-op Daily Weight Gain (gms.)	-.2**	.4	.6	.8
Av. Post-op Daily Weight Gain (gms.)	5.1	3.6	0	1.6
Av. % Increase in Food Consumption	88%	53%	5%	36%
Av. % Increase in Body Weight*	23%	11%	-1%	1%

* The formula <u>FINAL WT. - PROJECTED WT.</u> x 100 was used. The projected weight was based on the
                 PROJECTED WT.
average pre-operative weight gain.

** The projected weight was calculated by the more conservative assumption of no weight gain
rather than a loss.

25

absent and replaced by a median raphe. The general appearance was very much like that of a male. Starting at approximately 100 days of age these animals were tested weekly for several years in peer groups containing four to six male and female monkeys of similar size and age (31). Prior to this they had been left undisturbed with their mothers. A great variety of behavioral elements were scored during the tests, but particular emphasis was given to that behavior which Dr. Leonard Rosenblum (87), working in Harlow's laboratory (45), had shown differed in young male and female monkeys. In general, the young males are

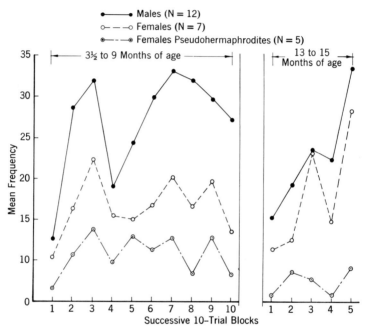

Figure 12. Frequency of "rough and tumble" play during the first and second years of life. [After Goy (31).]

more forceful in their play patterns while the females are more passive. Figure 12 illustrates the amount of "rough and tumble" play exhibited by normal males and females and the genetic female pseudohermaphrodites. In the left-hand graph the behavior from 3½ to 9 months is shown; the right-hand graph shows the results of the tests given to animals from 13 to 15 months. It is evident that the masculinized females are displaying much more of the "rough and tumble" play than normal females.

Figure 13 illustrates the frequency of "chasing play" displayed during this same period by these animals. The masculinized females during the 13- to 15-month testing period exhibited as much "chasing play" behavior as the normal males, and both of these groups are significantly higher in this measure than the normal females. Similar results have been obtained with "gape" behavior, sometimes called "social threat," which consists of an abrupt mouth opening with teeth covered directed at another animal. The normal males and the masculinized genetic females exhibited significantly more of this behavior than did normal females.

During the first year of life male monkeys display a fair amount of mounting behavior during these group testing sessions. Table 4 summarizes the mounting

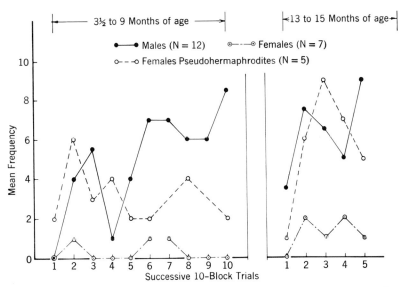

Figure 13. Frequency of "chasing play" during the first and second years of life. [After Goy (31).]

exhibited by all members of the three groups tested to date. Whereas the normal female hardly mounted at all (the median score is zero), the pseudohermaphroditic females mounted almost as much as the normal males. It is possible that the "rough and tumble" and "chasing play" as well as the mounting are all related to sexual behavior. The normal male may have to develop a certain amount of aggressiveness for sexual behavior to be effective. In any case these masculinized females have been changed in dramatic and basic ways and these changes by themselves might be expected to precipitate a number of other behavioral developments in this very social animal. These animals are still being studied and it will be extremely interesting to learn of the later developments in social behavior as well as sexual behavior.

TABLE 4. TOTAL NUMBER OF MOUNTS DISPLAYED IN 100 DAILY OBSERVATIONS DURING THE FIRST YEAR OF LIFE

Normal Males		Pseudohermaphroditic Females		Normal Females	
Animal No.	Total Frequency of Mounting	Animal No.	Total Frequency of Mounting	Animal No.	Total Frequency of Mounting
839	19	828	20	830	0
1242	117	829	7	831	0
1243	169	836*	4	844	0
1617	44	1239	9	845	0
1620	11	1616	9	847	0
1625	4	1619	22	848	0
1636	0	1640	33	1252	1
1657	5	1656	0	1551	1
1658	12	Median	9.0	1642	0
1662	1			1649	0
1954	18			1654	0
1958	28			1769	0
2331	11			1838	0
2332	19			2320	0
2333	9			2334	0
2354	4			2344	0
2356	12			2349	0
2358	2			2350	0
2359	15			2362	1
2552	8			2369	0
2555	3			2539	0
2557	25			2551	0
Median	11.5			2569	0
				2575	0
				2577	0
				2580	0
				Median	0

*Data obtained between 10-1/2 and 16 months of age rather than 3-1/2 to 9 months as in all other cases.

Courtesy of Dr. Robert W. Goy.

The general conclusion we would draw from this survey is that an increasing amount of evidence indicates that steroid hormones exert a different influence on the nervous system at different stages of development. At an early undifferentiated stage androgen may have an organizational influence. Harris (48) prefers the term "inductive" in this context; Barraclough (3) refers to the action of androgen at this time as "steroid imprinting." Whatever term is employed, the suggestion is that the nervous system exposed to androgen during a critical period will be permanently altered. The capacity to control the anterior pituitary gonadotrophic activity in a cyclical pattern will be lost and with respect to behavior the nervous system will be "tuned" in some way to change the probabilities of different behavioral elements being expressed. At a later age the steroid hormones serve as activators of the substrate already organized.

It has been pointed out that there is an evolutionary trend in the action of steroid hormones on behavior (7). In general, there seems to be an emergence from dependence on these hormones. Animals lower in the phylogenetic scale exhibit very little or no sexual behavior following gonadectomy, whereas the sexual behavior of animals at higher levels may persist for long periods without gonadal hormones. It should be noted, however, that the evidence for man is not this clear: Money (74) has described a number of cases of hypogonadal males in which sexual libido seemed to be very dependent upon androgen therapy. The human data, however, are quite variable. Nevertheless, a great amount of evidence can be mustered in support of this evolutionary trend toward emancipation from dependence on the activational role of steroid hormones. This is particularly evident in the case of the female, where only in the lower mammals is sexual behavior completely dependent upon ovarian hormones. The sexual behavior of higher mammal females is more thoroughly integrated with social behavior such as that involved in dominance-submissive patterns, and as a result its persistence is difficult to interpret. In addition, in the case of males most of the evidence illustrating persistence of sexual behavior without steroid hormones has been obtained from postpubertally gonadectomized, "experienced" animals (86). More relevant in the present context are the recent findings of the organizational impact of steroid hormones in primates, which raise the question of whether the evolutionary trend applies equally to the activational and organizational actions of these hormones.

There is simply not enough information at this time to be very specific about the mechanisms underlying the steroid hormone induced changes, which are being called organizational. The evidence that the changes are in the nervous system and perhaps especially in the hypothalamus seems reasonably convincing. The major role of the hypothalamus is supported by de Groot's (38) preliminary finding that even in the case of a surgically isolated hypothalamus cyclical functioning of pituitary gonadotrophins may continue. The problem of critical localization within the hypothalamus has been investigated by Halász and Gorski (42).

In previous work Halász and his associates (43, 44) have shown that the medial basal hypothalamic area (MBH), which includes the arcuate and periventricular nuclei and the ventromedial medial region posterior to the optic chiasma, is the only area capable of maintaining the structure and function of the transplanted pituitary. Pituitary transplants located in other hypothalamic areas were not maintained. This critical hypothalamic area has been referred to by these workers as the "hypophysiotrophic area."

Electrolytic lesions only in the MBH consistently produce gonadal atrophy. By utilizing a fine rotating knife it was possible to perform a series of selective deafferentations of the MBH. If the MBH was completely separated from the rest of the hypothalamus, ovulation did not occur and the normal ovarian compensatory hypertrophy seen after removal of one ovary did not occur. If the MBH was isolated from all of the hypothalamus except the anterior region, normal anterior pituitary-ovarian functioning was present. Gonadotrophic secretion appeared to be reasonably well maintained in males in which the MBH was completely isolated, and under the same conditions ovarian atrophy was not seen. The authors therefore conclude that tonic gonadotrophic release can be maintained by the MBH alone, but connections with the anterior hypothalamus are necessary for cyclicity.

In the present context it does not seem unreasonable to entertain the hypothesis that one of the functions of early exposure to androgen is to disrupt this connection between the anterior hypothalamus and the MBH. Consistent with this conclusion—early androgen disrupts hypothalamic functioning in the female— are the data obtained demonstrating that female rats, sterilized by early androgen treatment, are capable of displaying normal gonadotrophic functioning in response to electrical stimulation of appropriate hypothalamic areas. Barraclough (3) concluded that androgen treatment was responsible for creating a neuronal refractoriness to the exteroceptive stimuli such as light and interoceptive stimuli such as steroids responsible for the cyclic discharge of gonadotrophins which produce ovulation. Consistent with the conclusion drawn from the hypothalamic deafferentation studies is the evidence presented by Barraclough (3) that the preoptic area is the primary area changed by early testosterone administration, but with higher dosages of testosterone the arcuate-ventromedial nuclear complex may be involved as well. It should be clear that this discussion of the relationship of the anterior hypothalamus to the MBH has concentrated on the cyclicity in gonadotrophic hormone release. The neural substrate that lies at the basis of the sexual differences in behavior is undoubtedly much more complex.

It is conceivable that the cyclical secretion of gonadotrophins is entirely regulated within the hypothalamus and does not depend on feedback from ovarian hormones. Recent work by Inoué (56) makes this possibility appear unlikely. This investigator transplanted ovaries into the liver of ovariectomized animals. Because the liver destroys most of the estrogen secreted by the ovary under the conditions of an intrasplenic transplant, it was possible to evaluate the significance of estrogen feedback. Under these conditions cyclicity was disrupted. If additional ovarian tissue was transplanted elsewhere in this preparation, cyclicity was reestablished.

The recent report by Roberts, Steinberg, and Means (85) that electrical stimulation of hypothalamic areas in the opossum produces vigorous display of complex male mating patterns in the female as readily as the male is certainly relevant to any speculation about the nature of the organizational influence on the neural substrate mediating behavior. If electrical stimulation can be said to be activating organized neural patterns and it is possible to generalize from these results with the opossum, it would seem that early androgen does not selectively establish motor connections within the nervous system. Both motor patterns seem to exist in both sexes. In the work with the opossum the intensity of stimulation necessary to elicit male sexual behavior in females was not higher than that required when males were the subjects. Nor did the response seem forced or stereotyped, since sexual behavior was not displayed unless an appropriate stimulus object was present. The conclusion drawn by Roberts and his co-workers is that early androgen exerts its influence by determining later "sensitivity of the mechanism to hormones or the development of neural input" (85, p. 14).

The possibility that early administration of androgen may modify the motivational significance of sensory input has to be considered in addition to its influence on the future sensitivity to hormones. The studies of Goy, Phoenix, and Young referred to earlier are relevant here. These workers have demonstrated the modification of the play patterns of "masculinized" female monkeys observed during the first 1½-years of life. These animals responded essentially as males. Recently, Dr. John Resko (83), working in the same laboratory with the relatively specific chemical method of gas chromatography with electron capture, has shown that even large samples of plasma from monkeys between 6 months and 1½-years of age do not contain measurable amounts of testosterone. It would seem unlikely, then, that these masculine play patterns are being activated by androgen at this young age. It is more likely that the various stimuli present in the group testing situation have different motivational significance for males and females or a different capacity to elicit specific responses.

An analogy that may help to concretize this hypothesis may be found in the micturitional pattern of dogs (11, 69). The three main differences between the micturition behavior of the adult male and bitch concerns the leg lifting versus squatting, "piecemeal" versus "all or none" urination, and the orientation toward vertical objects versus relatively indiscriminate selection of locations. It is clear that androgen plays an active role in the development of this pattern. Castration of the adult male results in loss of leg lifting and replacement androgen therapy reestablishes leg lifting. Androgen administration did not produce leg lifting in the adult bitch, but it did cause her to adopt the male pattern of depositing only a few drops at a time. If androgen was administered to the female puppy, however, not only did she develop the leg lifting pattern, but she directed the urine to conspicuous, vertical objects. For the normal male a combination of interoceptive cues (visceral feedback from bladder, etc.) and the presence of an exteroceptive stimulus with motivational significance (fire hydrant) produces a specific behavioral orientation. These data should be replicated with a greater number of animals,* but if it is true that early androgen administration changes the genetic female dog's orientation to vertical objects, this would suggest that one of the changes produced by the steroid hormone is the modification of the motivational valence of sensory input. It would, however, still be a major challenge to translate this concept into physiological mechanisms. A preliminary step in this direction was attempted by Valenstein (93), who suggested that the motivational significance of sensory input may be determined by the extent to which reinforcing neural structures are activated. According to this view, drive states and hormonal levels act as a "gating mechanism," directing afferent impulses and thereby determining their ability to activate those neural areas critical to the reinforcement process.

In presenting a survey of our changing view of the action of steroid hormones on the nervous system and behavior, it has become apparent that it will not be possible to cover the field completely. Some relevant topics have been omitted entirely and in some instances difficult problems have not been presented with adequate attention to the complexity of the issue. It is to be hoped that no gross distortion has been introduced in attempting to provide an overview. The recent rate of progress in this field has been most rapid and a number of problems should be solved during the next few decades. It would be delighted to have the opportunity of presenting another survey at the University of Michigan's bicentennial celebration.

REFERENCES

1. Barfield, R. J. Induction of copulatory behavior by intracranial placement of androgens in capons. *Amer. Zool.,* **4**:301, 1964.
2. Barraclough, C. A. Production in anovulatory, sterile rats by single injections of testosterone propionate. *Endocrinology,* **68**:62-67, 1961.
3. Barraclough, C. A. Modification in reproductive function after exposure to hormones during the prenatal and early postnatal period. In L. Martini and W. F. Ganong. (Eds.), *Neuroendocrinology,* New York: Academic Press, 1967, Vol. II, pp. 61-99.
4. Barraclough, C. A., and Gorski, R. A. Evidence that the hypothalamus is responsible for androgen induced sterility in the female rat. *Endocrinology,* **68**:68-79, 1961.

---

*At the Field Station for Research on Animal Behavior at the University of California (Berkeley), Professor Frank A. Beach is currently studying the micturition pattern of male and female beagles that have received hormonal manipulations at different ages. Preliminary results suggest that the relationships may be more variable than originally described.

5. Beach, F. A. Comparison of copulatory behavior of male rats raised in isolation, cohabitation and segregation. *J. Genet. Psychol.,* **60**:121-136, 1942.

6. Beach, F. A. Hormonal induction of mating responses in a rat with congenital absence of gonadal tissue. *Anat. Rec.,* **92**:289-292, 1945.

7. Beach, F. A. Sexual behavior in animals and men. *The Harvey Lectures 1947-48.* Springfield, Ill.: Charles Thomas, 1950, pp. 254-280.

8. Beach, F. A. Normal sexual behavior in male rats isolated at fourteen days of age. *J. comp. physiol. Psychol.,* **51**:37-38, 1958.

9. Beach, F. A., and Holz-Tucker, A. M. Effects of different concentrations of androgen upon sexual behavior in castrated male rats. *J. comp. physiol. Psychol.,* **42**:433-453, 1949.

10. Beach, F. A., and Levinson, G. Effects of androgen on the glans penis and mating behavior of castrated male rats. *J. exper. Zool.,* **114**:159-171, 1950.

11. Berg, I. A. Development of behavior: the micturition pattern in the dog. *J. exper. Psychol.,* **34**:343-368, 1944.

12. Broadhurst, P. L. Determinants of emotionality in the rat: I. Situational factors. *Br. J. Psychol.,* **58**:1-12, 1957.

13. Burns, R. K. Hormones and the differentiation of sex. In G. S. Avery, Jr., et al. (Eds.), *Survey of Biological Progress,* New York: Academic Press, 1949, Vol. I, pp. 233-266.

14. Burns, R. K. Role of hormones in the differentiation of sex. In W. C. Young (Ed.), *Sex and Internal Secretions,* Baltimore, Md.: Williams and Wilkins, 1961, third ed., pp. 76-158.

15. Conner, R. L., Levine, S., Wertheim, G. A., and Cummer, J. F. Hormonal determinants of aggressive behavior. In *Experimental Approaches to the Study of Emotional Behavior,* New York: Annals of New York Academy of Science, 1968, in press.

16. Dantchakoff, V. Rôle des hormones dans la manifestation des instincts sexuals. *Compt. rend. Acad. Sc.,* **206**:945-947, 1938.

17. Dantchakoff, V. Sur les effects de l'hormone male dans une jeune cobaye femelle traité depuis un stade embryonnaire (inversions sexuelles). *Comp. rend. Soc. biol.,* **127**:1255-1258, 1938.

18. Dantchakoff, V. Sur les effets de l'hormone male dans un jeune cobaye male traité depuis un stade embryonnaire (production d'hypermales). *Comp. rend. Soc. biol.,* **127**:1259-1262, 1938.

19. Dantchakoff, V. Actions de la testostérone sur les énergies globales de l'organisme. *Comp. rend. Soc. biol.,* **141**:114-116, 1947.

20. Donovan, B. T. The regulation of the secretion of follicle-stimulating hormone. In G. W. Harris and B. T. Donovan (Eds.), *The Pituitary Gland,* Berkeley: University of California Press, 1966, Vol. 2, pp. 49-98.

21. Everett, J. W. Central neural control of reproductive functions of the adenohypophysis. *Physiol. Rev.,* **44**:373-431, 1964.

22. Everett, J. W., Sawyer, C. H., and Markee, J. E. A neurogenic timing factor in control of the ovulatory discharge of luteinizing hormone in the cyclic rat. *Endocrinology,* **44**:234-250, 1949.

23. Fisher, A. E. Maternal and sexual behavior induced by intracranial chemical stimulation. *Science,* **124**:228-229, 1956.

24. Flerkó, B. Control of gonadotropic secretion in the female. In L. Martini and W. F. Ganong. (Eds.), *Neuroendocrinology,* New York: Academic Press, 1966, p. 15.

25. Flerkó, B., and Szentágothai, J. Oestrogen sensitive nervous structures in the hypothalamus. *Acta Endocr.,* **26**:121-127, 1957.

26. Gerall, A. A., and Ward, I. I. Effects of prenatal exogenous androgen on the sexual behavior of the female albino rat. *J. comp. physiol. Psychol.,* **62**:370-375, 1966.

27. Gollin, E. S. Reversal learning and conditional discrimination in children. *J. comp. physiol. Psychol.,* **58**:441-445, 1964.

28. Gollin, E. S. Conditions which facilitate or impede cognitive functioning: Implications for developmental theory and for education. In R. D. Hess and R. M. Bear (Eds.), *Early Education: Current Theory, Research, and Practice,* Chicago: Aldine Press, 1968.

29. Gollin, E. S., and Liss, P. Conditional discrimination in children. *J. comp. physiol. Psychol.,* **55**:850-855, 1962.

30. Gorski, R. A., and Barraclough, C. A. Effects of low dosages of androgen on the differentiation of hypothalamic regulatory control of ovulation in the rat. *Endocrinology,* **73**:210-216, 1963.

31. Goy, R. W. Organizing effects of androgen on the behavior of rhesus monkeys. In R. P. Michael (Ed.), *Proceedings of the London Conference: Endocrines and Human Behavior,* in press.

32. Goy, R. W., Bridson, W. E., and Young, W. C. Period of maximal susceptibility of the prenatal female guinea pig to masculinizing action of testosterone propionate. *J. comp. physiol. Psychol.,* **57**:166-174, 1964.

33. Goy, R. W., and Young, W. C. Somatic basis of sexual behavior patterns in guinea pigs: factors involved in the determination of the character of the soma in the female. *Psychosom. Med.,* **19**:144-151, 1957.

34. Goy, R. W., and Young, W. C. Strain differences in the behavioral responses of female guinea pigs to alpha-estradiol benzoate and progesterone. *Behavior,* **10**:340-354, 1957.

35. Grady, K. L., Phoenix, C. H., and Young, W. C. Role of the developing rat testis in differentiation of the neural tissue mediating sexual behavior. *J. comp. physiol. Psychol.,* **59**:176-182, 1965.

36. Gray, J. A., Levine, S., and Broadhurst, P. L. Gonadal hormone injections in infancy and adult emotional behavior. *Animal Behav.,* **13**:33-45, 1965.

37. Green, J. D. The comparative anatomy of the hypophysis with special reference to its blood supply and innervation. *Amer. J. Anat.,* **88**:225-311, 1951.

38. Groot, J. de. In discussion: *Proceedings of the International Union of Physiological Sciences* XXII, International Congress, 1962, Vol. I, p. 623.

39. Grumbach, M. M., and Ducharme, J. R. The effects of androgens on fetal sexual behavior. *Fertility & Sterility,* **11**:157-180, 1960.

40. Grunt, J. A., and Young, W. C. Differential reactivity of individuals and the response of the male guinea pig to testosterone propionate. *Endocrinology,* **51**:237-248, 1952.

41. Grunt, J. A., and Young, W. C. Consistency of sexual behavior patterns in individual male guinea pigs following castration and androgen therapy. *J. comp. physiol. Psychol.,* **46**:138-144, 1953.

42. Halász, B., and Gorski, R. A. Gonadotrophic hormone secretion in female rats after partial or total interruption of neural afferents to the medial basal hypothalamus. *Endocrinology,* **80**:608-622, 1967.

43. Halász, B., Pupp, L., and Uhlarik, S. Hypophysiotrophic area in the hypothalamus. *J. Endocrinol.,* **25**:147-154, 1962.

44. Halász, B., Pupp, L., Uhlarik, S., and Tima, L. Further studies on the hormone secretion of the anterior pituitary transplanted into the hypophysiotrophic area of the rat hypothalamus. *Endocrinology,* **77**:343-355, 1965.

45. Harlow, H. Sexual behavior in the rhesus monkey. In F. A. Beach (Ed.), *Sex and Behavior,* New York: John Wiley and Sons, 1965, pp. 234-265.

46. Harris, G. W. Electrical stimulation of the hypothalamus and the mechanism of neural control of the adenohypophyses. *J. Physiol.* (London), **107**:418-429, 1937.

47. Harris, G. W. Central control of pituitary secretion. In J. Field (Ed.), *Handbook of Physiology,* Washington, D. C.: Amer. Physiological Soc., 1960, Sec. I, Vol. II, *Neurophysiology,* pp. 1007-1038.

48. Harris, G. W. Sex hormones, brain development and brain function. The Upjohn Lecture of the Endocrine Society. *Endocrinology,* **75**:627-648, 1964.

49. Harris, G. W., and Campbell, H. J. The regulation of the secretion of luteinizing hormone and ovulation. In G. W. Harris and B. T. Donovan (Eds.), *The Pituitary Gland,* Berkeley: University of California Press, 1966, Vol. II, *Anterior Pituitary,* pp. 99-165.

50. Harris, G. W., and Jacobsohn, D. Functional grafts of the anterior pituitary gland. *Proc. Roy. Soc.,* **B139**:263-276, 1952.

51. Harris, G. W., and Levine, S. Sexual differentiation of the brain and its experimental control. *Proc. Physiol. Soc., J. Physiol.,* 163:42-43P (Abst.), 1962.

52. Harris, G. W., Michael, R. P., and Scott, P. P. Neurological site of action of stilbesterol in eliciting sexual behavior. *Ciba Found. Symp., Neurol. Basis Behavior,* pp. 236-254, 1958.

53. Harris, G. W., and Michael, R. P. The activation of sexual behavior by hypothalamic implants of oestrogen. *J. Physiol.* (London), 171:275-301, 1964.

54. Haterius, H. O., and Derbyshire, A. J., Jr. Ovulation in the rabbit following upon stimulation of the hypothalamus. *Am. J. Physiol.,* 119:329-330, 1937.

55. Holweg, W., and Junkmann, K. Die hormonal-nervose regulierung der function des hypophysenborderlappens. *Klin. Worchschr.,* 11:321-323, 1932.

56. Inoué, S. Acyclic secretion of gonadotrophins by hypophysis in gonadectomized male and female rats. *J. Fac. Sc. Univ. Tokyo,* 9:309-318, 1961.

57. Kallas, H. Puberté précoce par parabiose. *C. R. Soc. Biol.* (Paris), 100:979-980, 1929.

58. Kallas, H. Developpement précoce de l'appareil génital chez le rat male infantile en parabiose. *C. R. Soc. Biol.* (Paris), 102:552-553, 1929.

59. Kallas, H. Parabiose und hypophysenvorderlappen. *Pflüg. Arch. ges. Physiol.,* 223:232-250, 1930.

60. Kennedy, G. C. Mating behavior and spontaneous activity in androgen sterilized female rats. *J. Physiol.* (London), 172:393-399, 1964.

61. Lehrman, D. S. The physiological basis of parental feeding in the ring dove (Streptopelia risoria). *Behavior,* 7:241-286, 1955.

62. Lehrman, D. S. Control of behavior cycles in reproduction. In W. Etkin (Ed.), *Social Behavior and Organization Among Vertebrates,* Chicago: University of Chicago Press, 1964, pp. 143-166.

63. Levine, S., and Mullins, R., Jr. Estrogen administered neonatally affects adult sexual behavior in male and female rats. *Science,* 144:185-187, 1964.

64. Levitt, R., and Webb, W. B. Effect of pentobarbital sodium on sleep latency and length of sleep in rat. *Nature,* 204:605-606, 1964.

65. Levitt, R., and Webb, W. B. Variables associated with a pentobarbital-induced sleep response. *Psychonomic Sci.,* 6:431-432, 1966.

66. Lisk, R. D. Diencephalic placement of estradiol and sexual receptivity in the female rat. *Am. J. Physiol.,* 203:493-496, 1962.

67. Marshall, F. H. A. Sexual periodicity and the causes which determine it. *Phil. Trans.,* B226:423-456, 1936.

68. Martinez, C., and Bittner, J. J. A non-hypophyseal sex difference in oestrus behavior of mice bearing pituitary grafts. *Proc. Soc. Exp. Biol.,* **91**:506-509, 1956.

69. Martins, T., and Valle, J. R. Hormonal regulation of the micturition behavior of the dog. *J. comp. physiol. Psychol.,* **41**:301-311, 1948.

70. Matthews, L. H. Visual stimulation and ovulation in pigeons. *Proc. Roy. Soc.,* **B126**:557-560, 1939.

71. McCann, S. M., and Dhariwal, A. P. S. Hypothalamic releasing factors and the neurovascular link between the brain and the anterior pituitary. In L. Martini and W. F. Ganong (Eds.), *Neuroendocrinology,* New York: Academic Press, 1966, pp. 261-289.

72. Michael, R. P. The selective accumulation of oestrogen in the neural and genital tissues of the cat. In *Proc. 1st Intern. Congr. Hormonal Steroids,* Milan, 1962. New York: Academic Press, 1964, Vol. II, pp. 457-469.

73. Michael, R. P. Oestrogens in the central nervous system. *Br. Med. Bull.,* **21**:87-90, 1965.

74. Money, J. Components of eroticism in man: The hormones in relation to sexual morphology and sexual desire. *J. Nerv. & Ment. Dis.,* **132**:239-248, 1961.

75. Moore, C. R. *Embryonic Sex Hormones and Sexual Differentiation. A Monograph in American Lectures in Endocrinology,* W. O. Thompson (Ed.). Springfield, Ill.: Charles Thomas, 1947.

76. Moore, C. R., and Price, D. Gonad hormone functions, and the reciprocal influence between gonads and hypophysis with its bearing on the problem of sex-hormone antagonism. *Am. J. Anat.,* **50**:13-71, 1932.

77. Neumann, F., and Elger, W. Physiological and psychical intersexuality of male rats by early treatment with an anti-androgenic agent (1, 2a-methylene-6-chloro-Δ^6-hydroxyprogesterone-acetate). *Acta endocr.* (Kbh.), Suppl. **100**:174, 1965.

78. Neumann, F., and Elger, W. Permanent changes in gonadal function and sexual behavior as a result of early feminization of male rats by treatment with an anti-androgenic steroid. *Endokrinologie,* **50**:209-225, 1966.

79. Neumann, F., and Elger, W. Steroidal stimulation of mammary glands in prenatally feminized male rats. *Europ. Pharm. J.,* **1**:120-123, 1967.

80. Pfeiffer, C. A. Sexual differences of the hypophyses and their determination by the gonads. *Am. J. Anat.,* **58**:195-226, 1936.

81. Phoenix, C. H., Goy, R. W., Gerall, A. A., and Young, W. C. Organizing action of prenatally administered testosterone propionate on the tissues mediating mating behavior in the female guinea pig. *Endocrinology,* **65**:369-382, 1959.

82. Phoenix, C. H., Goy, R. W., and Young, W. C. Sexual behavior: General aspects. In L. Martini and W. F. Ganong (Eds.), *Neuroendocrinology,* New York: Academic Press, 1967, Vol. II, pp. 163-196.

83. Resko, J. A. Plasma androgen levels of the rhesus monkey: The effects of age and season. *Endocrinology,* in press.
84. Riss, W., Valenstein, E. S., Sinks, J., and Young, W. C. Development of sexual behavior in male guinea pigs from genetically different stocks under controlled conditions of androgen treatment and caging. *Endocrinology,* 57:139-146, 1955.
85. Roberts, W. W., Steinberg, M. L., and Means, L. W. Hypothalamic mechanisms for sexual, aggressive, and other motivational behaviors in the opossum, Didelphis Virginiana. *J. comp. physiol. Psychol.,* 64:1-15, 1967.
86. Rosenblatt, J. S., and Aronson, L. R. The decline of sexual behavior in male cats after castration with special reference to the role of prior sexual experience. *Behavior,* 12:285-338, 1958.
87. Rosenblum, L. The development of social behavior in the rhesus monkey. Unpublished Ph.D. dissertation, 1961, Univ. of Wisconsin Libraries, Madison, Wisconsin.
88. Sawyer, C. H., Everett, J. W., and Markee, J. E. A neural factor in the mechanism by which estrogen induces the release of luteinizing hormone in the rat. *Endocrinology,* 44:218-233, 1949.
89. Scharrer, E., and Scharrer, B. *Neuroendocrinology.* New York: Columbia University Press, 1963.
90. Steinbaum, E. A., and Miller, N. E. Obesity from eating elicited by daily stimulation of the hypothalamus. *Am. J. Physiol.,* 208:1-5, 1965.
91. Swanson, H. H. Modification of sex differences in open field and emergence behavior of hamsters by neonatal injections of testosterone propionate. *J. Endocrinol.,* 34:6-7, 1966.
92. Swanson, H. H. Alteration of sex-typical behavior of hamsters in open field and emergency tests by neonatal administration of androgen or oestrogen. *Anim. Behav.,* 15:209-216, 1967.
93. Valenstein, E. S. The anatomical locus of reinforcement. In E. Stellar and J. M. Sprague (Eds.), *Progress in Physiological Psychology,* New York: Academic Press, 1966, Vol. I, pp. 149-190.
94. Valenstein, E. S., Cox, V. C., and Kakolewski, J. W. Sex differences in taste preference for glucose and saccharin solutions. *Science,* 156:942-943, 1967.
95. Valenstein, E. S., Cox, V. C., and Kakolewski, J. W. Sex differences in hyperphagia and body weight following hypothalamic damage. In P. J. Morgane and M. J. Wayner (Eds.), *Neural Regulation of Food and Water Intake,* New York: Annals of New York Academy of Science, in press.
96. Valenstein, E. S., Riss, W., and Young, W. C. Experimental and genetic factors in the organization of sexual behavior in male guinea pigs. *J. comp. physiol. Psychol.,* 48:397-403, 1955.

97. Valenstein, E. S., and Young, W. C. An experiential factor influencing the effectiveness of testosterone propionate in eliciting sexual behavior in male guinea pigs. *Endocrinology,* **56**:173-177, 1955.

98. Wells, L. J., and van Wagenen, G. Androgen-induced female pseudohermaphroditism in the monkey (Macaca mulatta): Anatomy of the reproductive organs. Carnegie Inst. Wash., Publ. No. 235, *Contribution to Embryol.,* **35**:95-106, 1954.

99. Whalen, R. E., and Nadler, R. D. Suppression of the development of female mating behavior by estrogen administered in infancy. *Science,* **141**:273-274, 1963.

100. Yazaki, I. Further studies on endocrine activity of subcutaneous ovarian grafts in male rats by daily examination of smears from vaginal grafts. *Annot. Zool. Jap.,* **33**:217-225, 1960.

101. Young, W. C. The hormones and mating behavior. In W. C. Young (Ed.), *Sex and Internal Secretions,* Baltimore, Md.: Williams and Wilkins, 1961, third ed., pp. 1173-1239.

102. Young, W. C., Dempsey, E. W., Myers, H. I., and Hagquist, C. W. The ovarian condition and sexual behavior in the female guinea pig. *Am. J. Anat.,* **63**:457-487, 1938.

103. Young, W. C., Goy, R. W., and Phoenix, C. H. Hormones and sexual behavior. *Science,* **143**:212-218, 1964.

104. Young, W. C., and Grunt, J. A. The pattern and measurement of sexual behavior in the male guinea pig. *J. comp. physiol. Psychol.,* **44**:492-500, 1951.

105. Young, W. C., Grunt, J. A., and Valenstein, E. S. Strength of sex drive during repeated tests of behavior in the male guinea pig. Proc. Am. Soc. Zool. *Anat. Rec.,* **111**:487, 1951.

106. Young, W. C., Myers, H. I., and Dempsey, E. W. Some data from a correlated anatomical, physiological and behavioristic study of the reproductive cycle in the female guinea pig. *Am. J. Physiol.,* **105**:393-398, 1933.

107. Zimbardo, P. G. The effects of early avoidance training and rearing conditions upon the sexual behavior of the male rat. *J. comp. physiol. Psychol.,* **51**:764-769, 1958.

BEHAVIORAL AND ANATOMICAL SEQUELAE OF

DAMAGE TO THE INFANT LIMBIC SYSTEM

ROBERT L. ISAACSON, ARTHUR J. NONNEMAN,
LEONARD W. SCHMALTZ

As will be amply documented in subsequent contributions of Kling and Tucker (Chap. IV) and Harlow and his co-workers (Chap. III), destruction of neocortical areas in the infant often tends to produce less debilitation than similar lesions in the adult. At least this holds true for some tasks under certain conditions of training and testing. However, little is known concerning the amount of sparing of function which would be expected from lesions of the brain below the neocortical mantle made during early periods of life. To this end we have been evaluating the anatomical and behavioral effects produced by destruction of the archicortex, the hippocampus, in animals at various stages of development.

We chose to work with animals subjected to lesions of the hippocampus for several reasons. Some reasons were of a quite practical nature, for example, our rather extensive work with animals with this type of lesion in the past. But there were more important reasons as well.

The hippocampus is a major portion of the "old brain" structures collectively termed the limbic lobe or the limbic system. Its anatomical relationships to other limbic areas and interconnections with diencephalic and midbrain systems are well known. Behaviorally, we now are able to recognize a number of situations in which hippocampal destruction make a "significant difference" (both statistically and in a real sense), although there is disagreement about how these differences should be interpreted [see Douglas (4) for a review]. For our experiments, we chose three types of behavioral problems that were likely to be of diagnostic value in terms of hippocampal destruction. These tasks were selected on the basis of relatively abundant evidence of impairments found in them following hippocampal destruction and because we thought it possible to adapt the techniques, when necessary, for use with cats. Also, we wanted tasks

The research supported in this chapter was supported, in part, by grant GB 533 OX from the National Science Foundation. We should like to express our appreciation to a number of people who have helped us in this work over the past four years. Those helping train the animals have been Miss Judith Plekker, Mrs. W. P. Archibald, Mr. H. Michael Peter, Mr. Jeffrey Fine and Mr. Dale Lewellyn. Mrs. Donald Zanotti and Mrs. Sven Lie have helped in the preparation of histological materials.

that would generate considerable amount of data covering a wide range of animal talents.

A word or two should be said about our choice of the cat for this research. In many ways the cat is less than ideal. Cats are often difficult to train, recalcitrant by nature, and subject to many common diseases which often prove fatal. For our present purposes they have a special advantage in that their size and vigor at birth is sufficient to allow them to recover from the surgical procedures. Some 80% or more of our animals subjected to surgery at birth survive the trauma and exhibit normal growth patterns. The operated animals are raised with their unoperated siblings, under the care of their natural cat mother. Actually, their postsurgical recovery is quite uneventful and no particular behavioral abnormalities have been noticed. Moreover, the neonatal brain of the cat is sufficiently well developed to allow reasonably accurate determination, during surgery, of the developing cortical and subcortical areas. This point was important since pilot studies with lower species made us disparage of achieving well localized lesions in the neonate.

About four years ago, when we began this work, we felt that the selection of the hippocampus was desirable because of its unique anatomical structure and appearance. The physical limits of the hippocampus can be readily observed once entry is gained to the lateral ventricles. The identification of boundaries of neocortical areas is quite difficult and subject to errors of evaluation and interpretation, especially in the neonate where development is rapidly progressing.

The hippocampus is readily identified both grossly and histologically. In the cat, even at birth, the alveus, which covers the outside of the hippocampus, is rather well developed and presents a shining, white outer surface as a glistening target during surgery. The characteristic histological appearance of the hippocampus was considered to be of value of ascertaining the extent to which our surgery was successful. The boundaries of neocortical areas are not as easily determined as the boundaries of the hippocampus and the dentate gyrus, which interlocks with it.

The cellular configuration of the hippocampus as it appears in cross section using a Nissl stain is shown in Figure 1, and a similar section prepared using the Heidenhain fiber stain is shown in Figure 2. Two sections through the brain of an infant cat showing the hippocampus as it appears in the neonate are presented in Figure 3 and 4.

We undertook our work believing that destruction of the hippocampus would serve our purposes because of the precision in surgery and lesion evaluation. As will be demonstrated shortly, our hope for extremely precise evaluation of the surgical damage which would be obtained by using the hippocampus, relative to neocortical areas, has not been entirely justified. This presents an interesting story in itself and merits considerable investigation in the future.

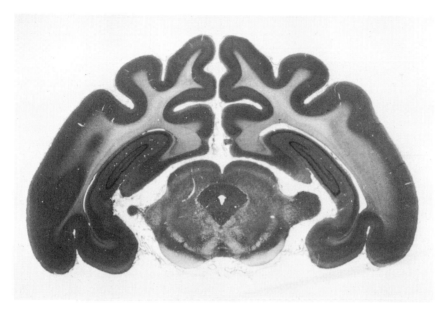

Figure 1. Thionin stained cross section of the adult cat brain showing the hippocampus at its fullest extent.

Our methods involve the direct removal of the hippocampus in cats of different ages. After a neonatal kitten has been anesthetized with 25 mg/kg of nembutal or with ether, the scalp is opened and the connective tissues scraped from the skull. At approximately the widest point, a small opening is made in the skull; this is extended both dorsally and ventrally. Then the dura is cut, the hippocampus is exposed by gentle aspiration of the neocortical surface, and an entrance is made into the lateral ventricles. Figure 5 shows the hippocampus through the defect produced in the skull and the cortex. In Figure 6 the same animal is shown with the head tilted at a slightly different angle to show the appearance of the brain through the defect after the hippocampus has been removed. The surface of the thalamus can be seen through the opening in this photograph. After hemostasis has been established, the wound is gently packed with Gelfoam and the temporal muscles pulled up over the cranial defect. Then the scalp is sutured and the animal returned as quickly as possible to its mother.

In the past three years. we have not brought any new animals into our cat colony and our lesions are made upon infants born in the laboratory. We have tried to achieve equal distribution of littermates among lesion groups.

Of course, not all of our animals survive surgery. Figure 7 shows the brain of a neonatal animal which did not survive; here the surgical defect can be

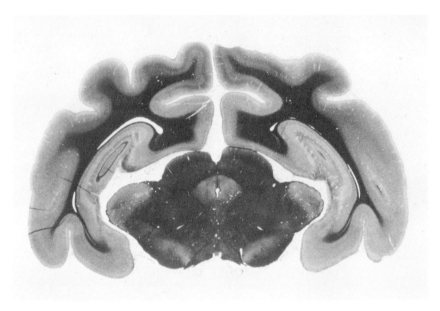

Figure 2. Heidenhain stained cross section of the adult cat brain at a level slightly posterior to that of Figure 1.

observed. In Figure 8 the intact side of this same animal is shown. The beginnings of a few of the major sulci can be seen, although the brain is not well convoluted at this early stage. This animal was intended to be a cortical control subject, and it is of interest to note that our survival rates have been lowest among animals with only neocortical destruction. Animals with both hippocampal and neocortical destruction tend to survive better than those which have only the neocortical damage. Others have reported difficulty in maintaining animals after neocortical destruction [e.g., Doty (3)].

ANATOMICAL SEQUELAE

One of our early and most striking observations, and one that was unexpected, was that animals which had received lesions of either the hippocampus or the cortex often had few signs of the damage when sacrificed in adulthood. Usually the animals remain in our laboratory some two or three years after the original surgery. Then they are sacrificed with an overdose of sodium pentobarbital and perfused with physiological saline, followed by 10% formalin. The brain is removed and subjected to our histological procedures.

From the beginning, we were struck by the rather limited amount of damage which was apparent during gross inspection of the brains of these animals. For

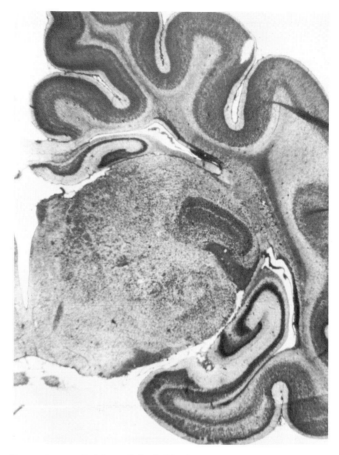

Figure 3. Photomicrograph (about 5x) of thionin stained cross section of one hemisphere of infant cat brain showing dorsal and ventrical portions of the hippocampus.

example, in Figure 9 we present some pictures of the top, right, and left sides of cat number 5 (named McKeachie). This cat is probably an extreme example in the sense that it has the least amount of superficial destruction of any that we have seen.

Figure 10 is a picture of a littermate of cat 5 which evidences somewhat more destruction, although on the right side of the animal (at the bottom of this figure) the damaged area is somewhat enlarged by an artificial gouge made when we were removing the brain from the skull. But, even in this case, if one contrasts the size of the lesion made on the day of birth with the appearance of the adult brain, the observable consequences on the surface of the brain are rather small.

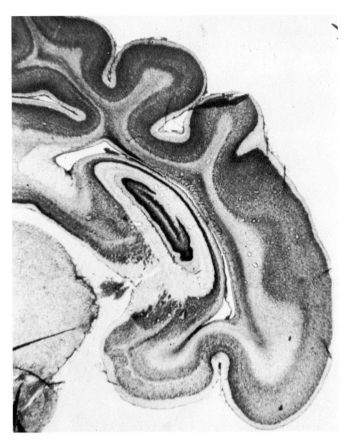

Figure 4. Photomicrograph (about 5x) of thionin stained cross section of one hemisphere of infant cat brain showing the hippocampus at a level posterior to that of Figure 3.

This "filling-in" of the wound produced by our surgical procedures is not restricted to animals with only hippocampal damage but occurs in animals with neocortical destruction as well. The amount of "filling-in" is not always complete and it may be that less "filling-in" takes place in animals with only cortical damage than in animals with both subcortical and cortical destruction. Figure 11 shows an animal which received neonatal destruction limited to the neocortex and, as can be seen, although there is some "filling-in," there is a considerable defect on the surface of the brain as well. Moreover, the lesion in this case produced more apparent distortions of the convolutions than has been seen in our earlier figures.

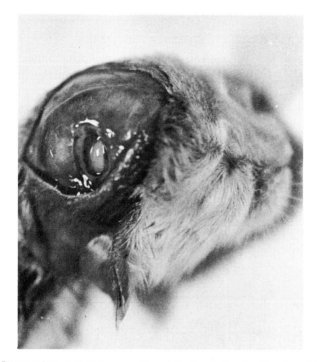

Figure 5. Photo taken during surgery of an infant cat showing the size of the defect in the skull. The neocortex has been removed and the exposed hippocampus can be seen.

Another point should be made here. The "filling-in" effect is not restricted to animals that receive lesions on the day of birth. In Figure 12 we have a picture of an animal with a unilateral lesion of the hippocampus made at 6 weeks of age. There is less distortion of the sulci, although there certainly is some deformation in the posterior suprasylvian gyrus, an extra fold, and there are unusual miniature sulci surrounding the lesion.

Let us examine in detail the internal structure of the adult brains of animals with neonatal hippocampal damage. Figure 13 shows a cross section of the hippocampal area of cat 5 (McKeachie). Note that despite the severity of the hippocampal damage induced on the day of birth, two years before, the hippocampus still seems to cover the entire extent of the lateral ventricle. The cells of the area, however, are much misaligned and their distortion is clearly evident. There is one band of giant hippocampal cells which has apparently followed a peculiar course of development, separating itself from the hippocampus proper and following a course along the outer ventrical wall into neocortex. The cells bend with the unnatural sulcus as if in an attempt to complete a numeral 9.

Cells of the dentate gyrus in the middle of the hippocampal formation seem to be spraying off in a "V" with an open-end directed toward the disjointed outer segment of the hippocampus. The hippocampal formation in another animal is shown in Figure 14. The bottom portion of the hippocampal formation shows a dentate gyrus typical of a cross section through the brain at a level where only a ventral segment of the hippocampus should be in evidence. The

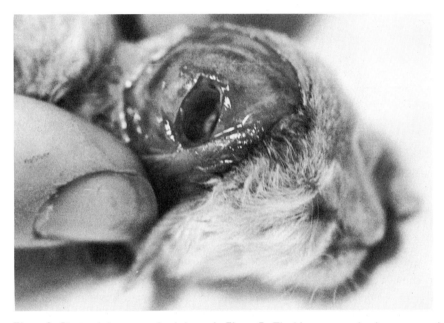

Figure 6. Photo of the same animal shown in Figure 5. The hippocampus has been removed and the dorsal surface of the thalamus can now be seen.

hippocampus extends dorsally until it ends, abruptly, in a mass of tissue which does not appear "hippocampal." The photomicrograph of the more dorsal regions of this section is presented in Figure 15.

Several things of interest can be noted in this figure. First, of course, is the abrupt transition between the ascending hippocampal tissue and the peculiar, rather amorphous tissue at its upper bound. Second is the abnormal tissue existing below and connected with the remaining dorsal hippocampus. What are we to call this tissue found below the reasonably well-defined dorsal hippocampus? It does not have the usual structural characteristics of hippocampal tissue.

Only a few small cells accept the Nissl stain in this region. Third, the peculiar portion of tissue between the abnormal dorsal and ventral hippocampal remnants has a line of cells which have the characteristics of hippocampal pyramidal cells, although the remaining tissue appears atypical of hippocampal material and is disconnected from the remaining dorsal and ventral aspects of hippocampus. This piece appears to have continuity with the lateral neocortical surface. These anomalies are the more surprising since the lesion was made at 3 months of age.

Often we find that the hippocampal tissue remaining after early destruction

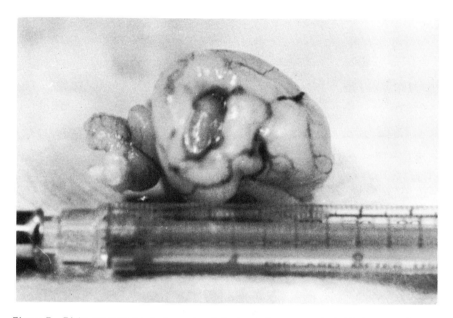

Figure 7. Right cerebral hemisphere of an infant cat after neocortical ablation exposing the hippocampus (1-cc syringe shown for comparison).

is directed toward the lateral surface, presumably toward the site of the original intrusion into the lateral ventricles. In Figure 16 the hippocampal tissue is connected into an area of white matter near the top of the lateral ventricle. The cellular configuration of hippocampus seems intact, although there appears to be a loss of cells in the dorsal, "inverted V" of pyramidal cells. Sometimes this cellular configuration is lost, although pyramidal cells are present, as is shown in Figure 17. Here dorsal and ventral remnants of hippocampus are directed toward the outer, neocortical mantle, and the normal configuration is almost completely absent. Even more interesting are the instances in which the residual

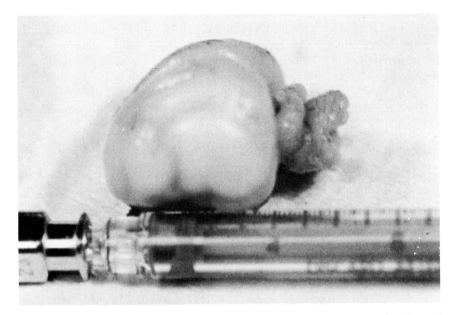

Figure 8.   Unoperated left cerebral hemisphere of same infant cat shown in Figure 7.
(1-cc syringe for comparison).

hippocampal tissue reaches toward the original neocortical defect and extends
to the outer surface of the brain. In Figure 18 the hippocampal tissue arising in
the middle left-hand side of the figure extends through a gap in the neocortex
and out to the surface.

Even though we personally created the extensive lesions shown earlier in the
neonatal animals, where we observed and often photographed the considerable
extent of hippocampal destruction, we were amazed to find such a considerable
amount of hippocampal tissue in the adult when sacrified two to four years
later. It would be tempting to consider the possibility of postnatal cell division,
yet there is considerable evidence that only for a few days the smaller cells of
the hippocampus undergo mitosis after birth (e.g., 1, 2) *in normal animals.* It
is possible that in lesioned animals inhibitory factors could be removed, allowing
for prolonged cell division or allowing cell division to be re-established after the
lesion. On the other hand, it is probably more likely that whatever hippocampal
tissue remains after surgery spreads out to fill the available space, following
lines of physical stresses and pressures.

At the present time we must defer judgment on the causes of these remarkable
histological results of the early lesions. They have stimulated us to undertake

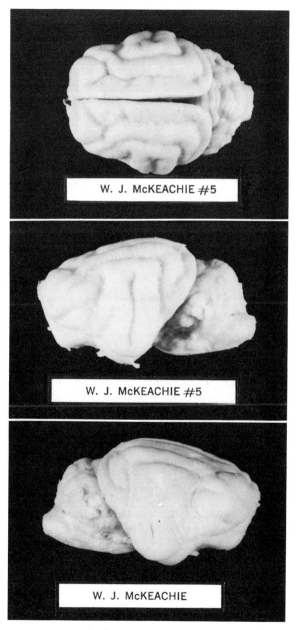

Figure 9. Dorsal, left, and right views of an adult cat brain (No. 5). This animal received bilateral hippocampal destruction as a neonate.

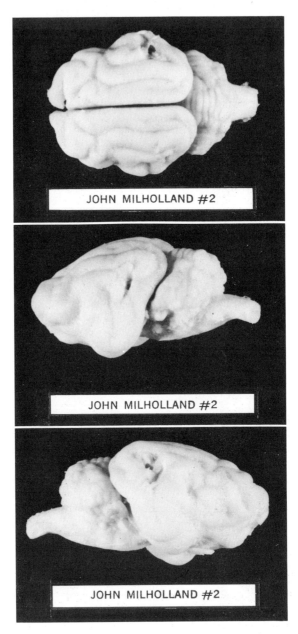

Figure 10. Dorsal, left, and right views of an adult cat brain (No. 2). This animal received bilateral hippocampal destruction as a neonate.

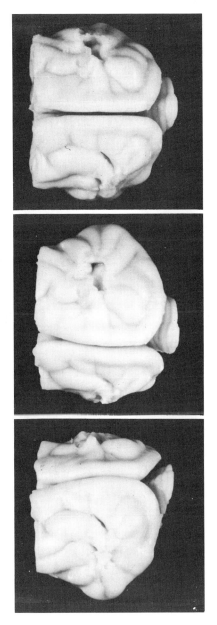

Figure 11. Left, right, and dorsal views of the posterior half of an adult cat brain (No. 28). This animal received bilateral neocortical destruction as an infant.

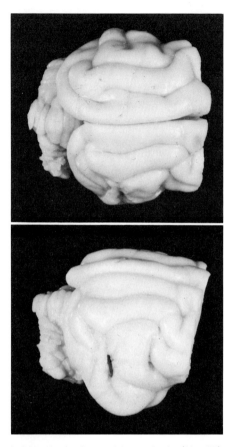

Figure 12.   Dorsal and right views of an adult cat brain (No. 65).   This animal received
unilateral neocortical destruction (right side) at six weeks of age.

new, strictly anatomical studies using unilateral lesions, and other work directed
using autoradiographic techniques to determine whether any significant cell division does occur following surgery.   But these heuristic values are offset by intro-
ducing a very practical difficulty in our present efforts.   Given the readjustive
mechanisms of the hippocampal tissues, it is impossible to use any known technique in describing quantitatively the amount of hippocampal tissue destroyed
by our lesions.   We can determine whether the lesion involved the hippocampus
or not, but more than this is difficult to support.

Considering the hippocampal changes produced by the early lesions is only
part of the story, however.   The gross "filling-in" of neocortical tissue which

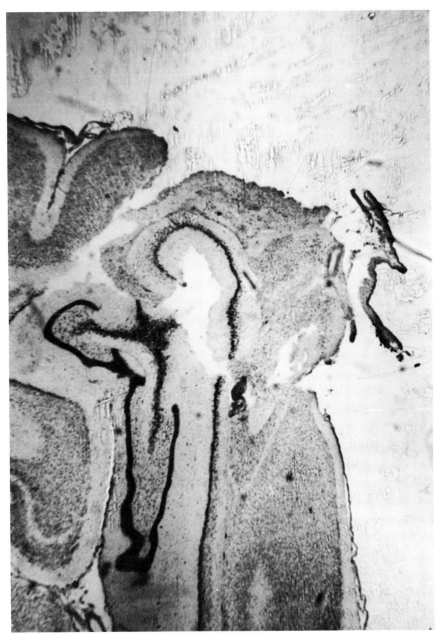

Figure 13. Photomicrograph (about 12.5x) of thionin stained cross section showing the hippocampal area of the brain pictured in Figure 9.

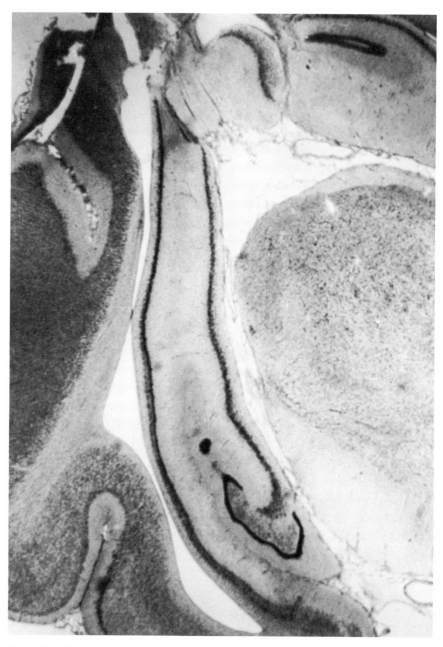

Figure 14. Photomicrograph (about 12.5x) of thionin stained cross section showing the hippocampal area of an adult cat which received bilateral hippocampal destruction at three months of age.

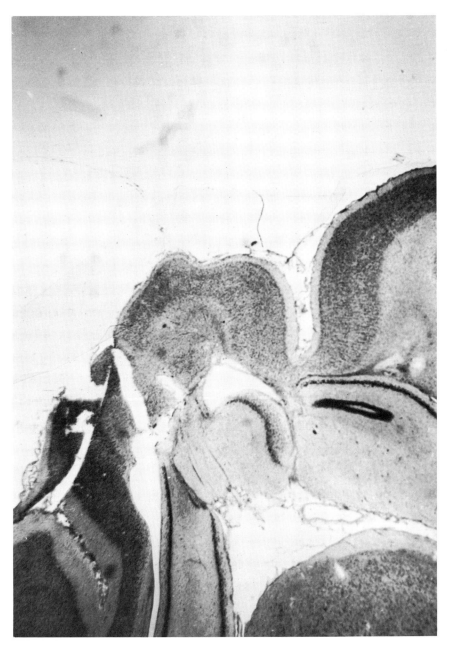

Figure 15. Photomicrograph (about 12.5x) showing the upper region of Figure 14.

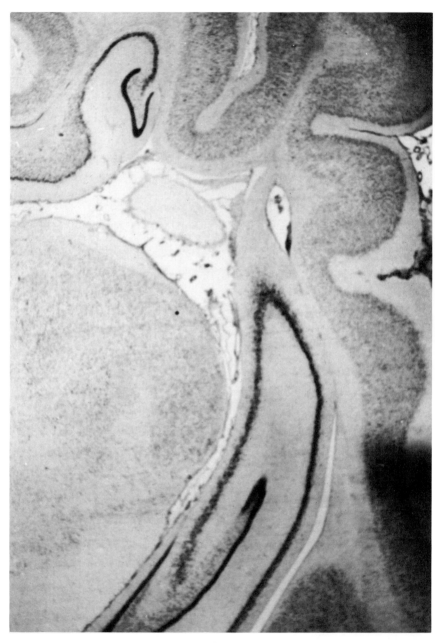

Figure 16. Photomicrograph (about 12.5x) of thionin stained cross section showing the hippocampal area of an adult cat which received bilateral hippocampal destruction as an infant.

Figure 17. Photomicrograph (about 50x) of thionin stained cross section showing peculiar hippocampal structure.

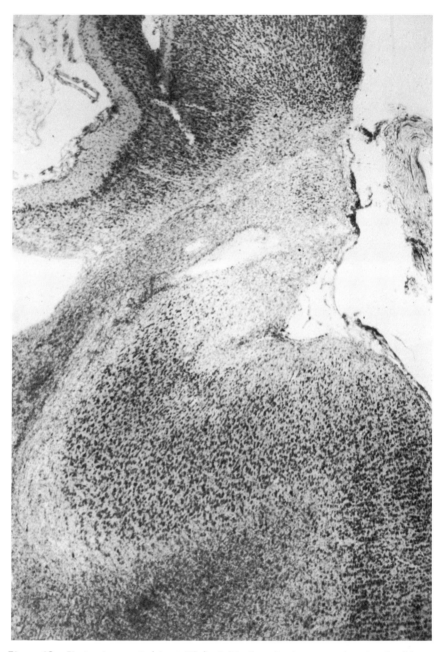

Figure 18. Photomicrograph (about 50x) of thionin stained cross section showing hippocampal tissue reaching the lateral neocortical surface.

occurs is not primarily of hippocampal origin. As shown before, the neonatal lesion involved a considerable amount of destruction of the neocortical surface. Yet, as noted, in the adult the defect is often small, always less than expected. Figure 19 shows a cross section through one of the gyri of a cat (Calico) which acted as a filler in the adult animal. The brain tissue of this unusual gyrus shows a considerable lack of laminar differentiation and only slight evidence of the usual "vertical organization" found in neocortex. Figure 20 shows the continuation of the gyrus shown in the previous figure and includes a relatively normal gyrus beneath. Signs of lamination and vertical organization are present in this lower gyrus.

Figure 21 shows the texture of the more dorsal and unusually structured gyrus of Figure 20 under higher magnification. The lack of cellular organization can be seen. Figure 22 reveals the usual texture of cortical areas under the same magnification.

At this time we are in a position only to note these anomalous conditions and report them without quantification. Studies of these anatomical abnormalities are underway, and we hope to be able to report them in the near future.

BEHAVIORAL SEQUELAE

As a general practice, the animals with lesions made early in life were allowed to recover for a period of at least 9 months before undergoing any formal training. After weaning, the animals were housed in our main cat colony in the laboratory, first with littermates and later in individual cages. They were allowed "free play" time daily and were handled and played with by those of us working in the laboratory.

After training was begun, usually at the age of 1 year, the animals continued with an experimental regime for another one to one and a half years. There was an interval of several months between each behavioral problem. The prolonged period of training illustrates one prime difficulty of using the cat as an experimental subject: cats frequently take excessive periods of time to master problems the rat or some other species might pick up in relatively short order.

The animals were observed during training on three types of behavioral tasks: a runway problem, a visual discrimination and reversal learning problem, and an operant conditioning task in which the animals had to master several DRL schedules. We shall take up these tasks separately, considering the performance of animals with the various brain lesions as well as the performance of intact animals.

For the purposes of this report we have excluded subjects in whom there were gross unintended neurological defects. These defects include unilateral or bilateral hydrocephalies or some degree of blindness due to accidental damage to the visual projection system. Also, we are including only cats for whom we have complete certainty as regards the lesion.

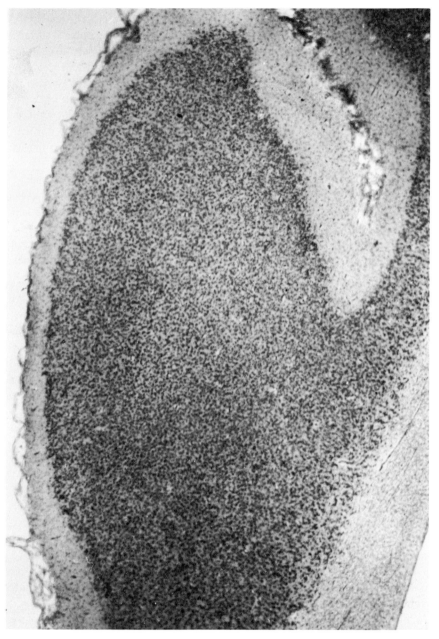

Figure 19. Photomicrograph (about 50x) of thionin stained cross section of an abnormal gyrus. See text for explanation.

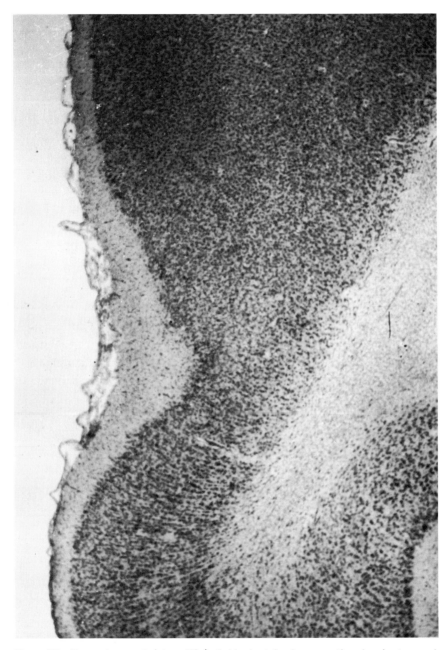

Figure 20. Photomicrograph (about 50x) of thionin stained cross section showing two gyri. The upper gyrus is the ventral continuation of the gyrus pictured in Figure 19.

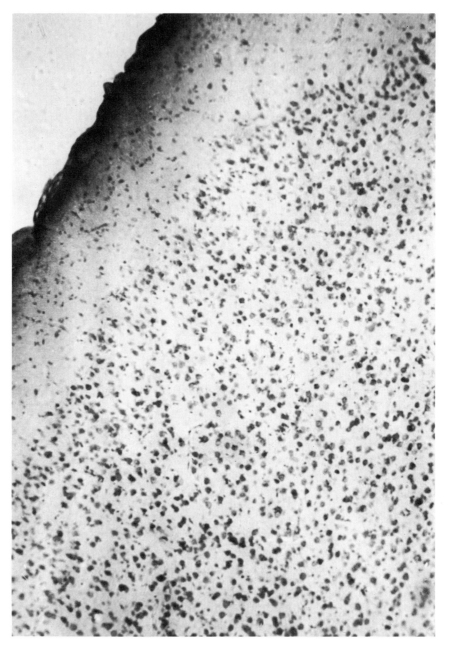

Figure 21. Photomicrograph (about 125x) of thionin stained cross section showing the upper gyrus of Figure 20 under higher magnification.

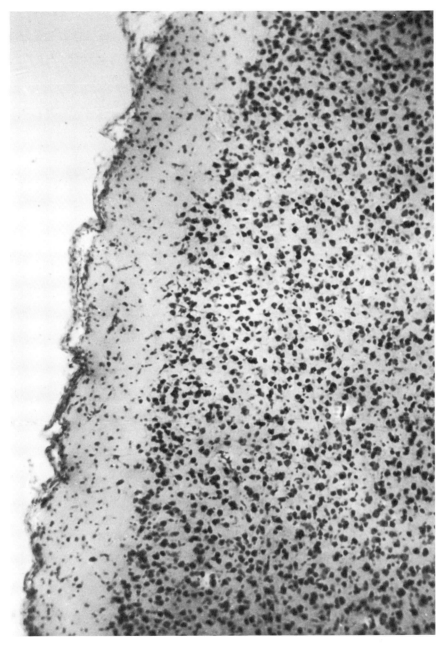

Figure 22. Photomicrograph (about 125x) of thionin stained cross section showing the lower gyrus of Figure 20 under higher magnification.

Since we make lesions in neonatal animals, and since littermates are often indistinguishable at birth, we sometimes have difficulty in determining the type of lesion made in animals until sacrifice. At times we feel this is carrying the principle of running the animals in a "blind manner" to an extreme, but it does provide another check on any possible bias we might accidentally introduce into the experiment.

All of the results that will be discussed are consistent with the data from a larger number of subjects, now in various stages of the experiments, which will be completing the series of tasks over the next year.

RUNWAY TEST

We have made considerable use of a special runway built for the cats. It is 6 feet long, 9 inches wide, and 15 inches high.

Animals are trained in this runway under 21 hours of food deprivation. Before training they are given experience with the runway for 4 days; during this time they experience prehandling and habituation to the runway itself for 10 minutes each day. In this pretraining period the animals are not given food in the runway. For 14 subsequent days the animals are trained to leave the start box, run to the goal box, and obtain a tuna fish reward. They are allowed to have commerce with, and eat, the tuna fish for 30 seconds. The intertrial interval is approximately 1 minute. Five trials are given each day.

Immediately after the animals have acquired this response, they are run in the apparatus for 8 consecutive days without food in the goal box. This is what we call the "extinction period." Afterward the rewards are reinstituted in the goal box and the animals trained for 7 more days. On the following day a "passive avoidance test" is given. The animals are given two normal trials and then two trials in which shock (2.5 ma) is delivered to the animals' feet in the goal box while they are eating the tuna fish. After the two shock trials, one more "normal trial" (without shock) is given on that day. For the next 4 to 5 days the training is conducted in a normal manner.

The data we anticipated would be of greatest interest are those from the extinction period and the passive avoidance test.

Figure 23 gives the time required for the animals to leave the start box on the first and eighth days of the extinction period. The solid bars on the histogram represent the latencies on the eighth day of extinction training and the shaded bars represent the latencies on the first day of the extinction period.

Animals with hippocampal destruction at either 6 weeks of age or as adults are not much affected by the omission of rewards at the end of the goal box. This supports observations in the rat (4, 5). The adult and 6-week lesion animals continue to leave the start box, even though the movement down the runway fails to produce any reward.

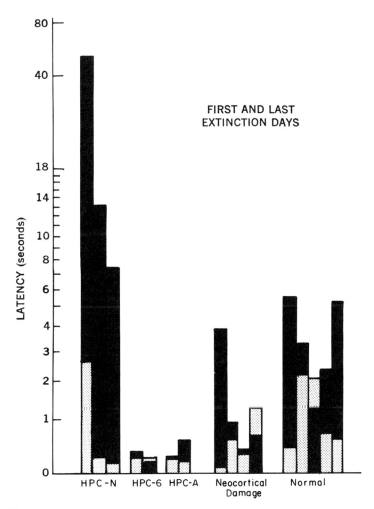

Figure 23. Histogram of exit latencies during extinction of the runway response. The shaded bars represent latencies on the first day of extinction; the solid bars represent latencies on the eighth and final day of the extinction period. (HPC-N, HPC-6, HPC-A; animals with hippocampal damage made as neonate, 6 weeks of age, and in adulthood, respectively.)

Contrasting with these animals is the performance of animals with neonatal hippocampal destruction, presented at the left of this bar graph. These animals tend to leave the start box somewhat more slowly than animals with only neocortical damage or normal animals. Thus we find in this task a remarkable differentiation between the effects of hippocampal destruction in neonates and 6 week old animals.

For the passive avoidance test the effect of shock on the runway performance of the animals after the 7 days of reacquisition training presents a rather different pattern. In Figure 24 all the animals with hippocampal lesions have been grouped together, as have animals with only the neocortical destruction. As can be seen, both types of brain damage seem to have some influence on passive avoidance behavior since all brain-damaged subjects leave the start box more readily than unoperated animals.

The animals to the left in the group with hippocampal damage are in fact animals with neonatal destruction. It would be easy to conclude that the production of hippocampal damage in the neonate tends to mitigate the effects of the impairment on the passive avoidance problem. However, this conclusion is not justified. A more detailed analysis of performance indicates that the neonatal hippocampally damaged animals are always more debilitated than animals with only neocortical damage or normal animals. We have discovered that there is an "experimenter bias" in the training of the animals such that some experimenters obtain generally shorter latencies from their subjects than do others. We have

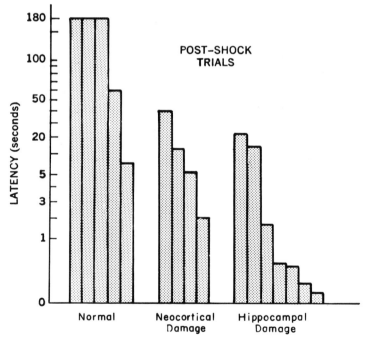

Figure 24. Histogram of exit latencies on the first postshock trial in the passive avoidance test.

analyzed the data in terms of the differences within littermates, siblings, and training groups. The most important variable for understanding the performance of animals appears to be the training groups to which an animal has been assigned. In some training groups we find generally shorter latencies and in others generally longer latencies. With this in mind, if we examine the effects of hippocampal lesions relative to neocortically damaged or intact animals, we find a significant decrease in the latencies on the postshock trials, which seems independent of the age at which the lesion was made. Without present data, it is impossible to conclude that making a hippocampal lesion at an early age produces a differential effect on this performance relative to adult lesions. What we can conclude is that the damage to the hippocampus always makes the exit latencies on the postshock trial less than cortical destruction alone or no destruction at all in the same training group.

It may be worthwhile to note that running speeds produce quite a different picture than do exit latencies from the start box. Figure 25 shows the effect produced by the shock on the postshock trial in terms of the time it takes the animal to get to the goal box after leaving the start box. As can be seen, the shortest running times are exhibited by animals suffering only neocortical destruction. The distinction between latencies and running times should be kept in mind when interpreting results from passive avoidance experiments. That is, a distinction should be made between the initiation of the act, which we feel is measured by exit latencies, and the tendency to continue this act until the goal is reached, which in this case would be measured by running times. Hippocampal destruction tends to affect the exit latencies, whereas neocortical lesions tend to affect running times.

OPERANT CONDITIONING PROGRAM

Next let us turn our attention to the changes in performance in the operant conditioning program, in particular, upon the DRL schedules. DRL schedules are schedules in which the animal must learn to wait a certain number of seconds between bar presses in order to obtain reinforcements. If a bar press is made at a shorter interval than permitted by the schedule, then the reinforcement is not provided and a timer is recycled so that a new interval is begun. If a DRL-10 schedule is being utilized, an animal must wait at least 10 seconds between responses. In a DRL-20 schedule an animal must wait 20 seconds between responses to obtain reinforcement.

Animals are trained under 21 hours of food deprivation, and dippers provide a milk reinforcement for appropriate responses. After shaping the animals to press the bar (which requires about 3 days), animals are placed on a continuous reinforcement schedule for 30 days. This long continuous reinforcement schedule is used to establish reasonable stability in the animals' bar press rates. Also, we wanted to build a solid performance on the CRF schedule so that the change

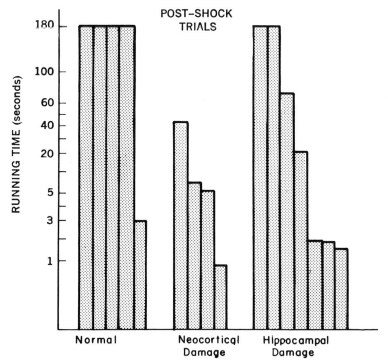

Figure 25. Histogram of running times of the first postshock trial in the passive avoidance test.

to the DRL procedures would be likely to produce the most dramatic effects. Schmaltz and Isaacson (7) have shown that the impairment exhibited by hippocampally damaged animals on the DRL-20 schedule arises when there is a transition from CRF to the delay schedule. If the animals have never experienced continuous reinforcement prior to the delay reinforcement schedule, then hippocampal destruction does not produce very great effects.

Animals were trained on a DRL-10 schedule to a criterion of 3 consecutive days of 50% or more reinforced responses (relative to the total number of responses emitted). After criterion was reached on the DRL-10 schedule, or after a maximum of 75 days, all animals were switched to the DRL-20 schedule. The animals were trained to the same criterion on the DRL-20 schedule, but we terminated this phase of the experiment at the end of 120 days of training if criterion performance had not reached before this time.

The bar graph given in Figure 26 shows the impairment usually found in animals with hippocampal destruction on the DRL schedules. These subjects take many more days to reach even low efficiency levels. In this figure we can observe that animals with neonatal destruction of hippocampus do *not* evidence this same impairment, and, as can be seen, their progression through successively more difficult criteria is about the same as that of normal subjects.

Comparisons among groups of animals with early lesions are possible from the data presented in Figure 27, which gives median response rates of the different groups of subjects. The anticipated increase in response rates found in animals receiving hippocampal damage as adults upon transition from continuous to intermittant reinforcement schedules is seen in our adult lesion group, where the median rate goes over 400 responses per half-hour session. Animals with hippocampal damage incurred at 3 months of age were indistinguishable from animals lesioned as adult. Animals with hippocampal damage sustained at 6 weeks of age approximated the performance of normal animals. Oddly enough, the performance of the neonatally damaged animals followed a different course of events. Their greatest increase in rate occurred at the transition to intermittent reinforcement schedules but did not diminish when the transition

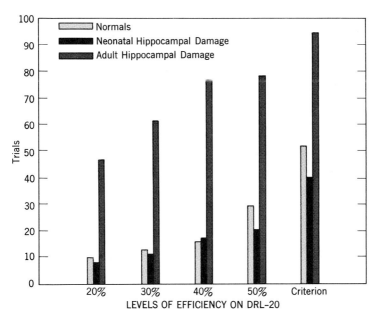

Figure 26. Histogram showing the mean number of days required to reach successive efficiency levels on the DRL-20 task. Criterion performance was 3 consecutive days at the 50% (or better) efficiency level.

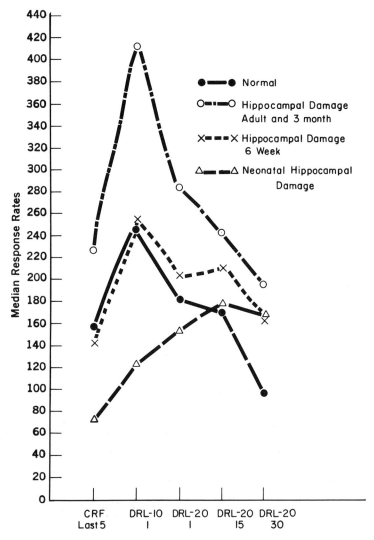

Figure 27. Median response rates at various stages during training on the operant conditioning task. The stage represented are on the abscissa: the last five days of continuous reinforcement, the first day of DRL-10 training, and the first, fifteenth, and thirtieth days of DRL-20 training.

from DRL-10 to DRL-20 occurred. Moreover, it increased over prolonged training on the DRL-20 schedule.

Thus, in terms of our results with the DRL schedules, we believe that destruction of the hippocampus made at an early age (6 weeks or before) produces behavioral effects which are substantially different from those produced by destruction at 3 months or beyond. Perhaps more important is the suggestion that the neonatally damaged subjects are substantially different from both normal animals and animals with later hippocampal destruction. This suggestion of qualitative difference in the hippocampally damaged neonate was found during extinction training in the runways as well.

DISCRIMINATION AND REVERSAL LEARNING TASK

In our discrimination learning and reversal task, we have used two experimental procedures which must be distinguished. In our first efforts two hutches were built and placed in a relatively large experimental room. Each hutch is 30 inches long, 18 inches wide, and 15 inches high. The cats were confined in a small start box to begin a trial and were allowed to progress freely toward the hutches when released from the start box. A black cloth was hung at the front of one hutch; a white cloth was hung at the front of the other. The hutches were located 5 feet in front of the start box and were 30 inches apart. The location of the white and black cloths were varied according to a Gellerman series. The animals had to learn to approach and enter the hutch with a black cloth covering the entrance for a reward, usually tuna fish, although sometimes a small piece of doughnut or liver was used. The food was obtained from a small food cup at the back of the hutch. Animals were trained with minimal or no deprivation for all of the acquisition period and for 24 days of reversal training. After this a 21-hour deprivation schedule was introduced. Three minutes were allowed between trials throughout training. We will refer to this procedure as experiment 1.

After gaining some experience with these procedures, we decided to make some changes in order to facilitate training in the future. Essentially, we made a wall flush with the entrances to the hutches. Rigid Polascreen doors, which could be illuminated from the rear, replaced the cloths at the hutch entrances. A gentle push on a door allows it to swing open and if the door is not the correct one, the correct one is automatically locked. This makes the training paradigm a "noncorrection" procedure. We will refer to this procedure as experiment 2.

Another procedural change made in experiment 2 was the reduction of the intertrial interval to 30 seconds.

In experiment 2 the plastic panel in front of one of the hutches was illuminated, whereas the other was not. The animals were trained to approach and enter the hutch whose front panel was not illuminated. Both problems were "brightness discrimination problems" in that animals approached the black (versus a white cloth) or the darker of the two plastic panels.

The animals were trained to a criterion of 9 or 10 correct responses in 10 trials for 3 consecutive days. When this performance level was reached, the reversal phase of the problem began; here the white cloth or brighter panel signalled the rewarded hutch.

Figure 28 shows the mean correct response scores during acquisition and reversal training. In experiment 1 there was some suggestion that the animals sustaining hippocampal damage as adults were slightly impaired in acquisition, but this difference was not found in experiment 2. It is doubtful whether much significance can be attributed to this result.

The animals with hippocampal destruction as adults were impaired in the reversal phase of this experiment (Figure 28b) as were animals with hippocampal destruction made at 6 weeks, or later, in experiment 2 (Figure 28d). The neonatally lesioned animals showed relatively normal reversal patterns in experiment 2.

A most significant fact presented in this figure is one we have only recently "discovered." This can be seen in the initial portions of the reversal curves. Even for the lesioned animals with impaired total performance, these curves rise to the 50% level as quickly as do the normal or cortical damaged Ss. This means that they are *not* perseverating the "old" correct response to the darker stimulus. They give up the old response as readily as do normal animals. This implies that an explanation of the performance of the brain damaged subjects on the basis of response perseveration is inadequate. Hippocampally damaged animals do not perseverate the old response, they give it up and reach a plateau of 50% correct responses due to a fixation of an approach response to one side or the other.

The impairment following hippocampal destruction in experiment 2 is caused by a fixation of response to one side. The adult lesioned animals of experiment 1 also showed this type of abnormal behavior, although their impairment was not as severe. Figure 29 shows data from the animals in experiment 1, plotted as histograms, in terms of trials to criterion, errors to criterion, and "days of fixation." During such days all responses are made to one hutch regardless of the pattern at the entrance. Figure 30 shows a similar analysis for the animals of experiment 2. Here the neonatally damaged animals show a tendency toward fixation, but it is by no means as severe as that found with animals suffering damage at older ages.

CONCLUSIONS

These results, taken as a whole, lead us to certain conclusions concerning the effect of destruction of the hippocampus at birth, in contrast to certain later ontogenetic periods.

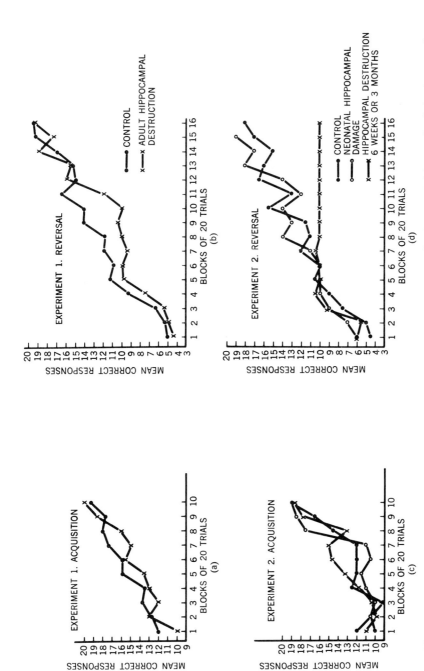

Figure 28. Mean correct response scores during acquisition and reversal training in the two simultaneous brightness discrimination experiments.

EXPERIMENT 1: REVERSAL

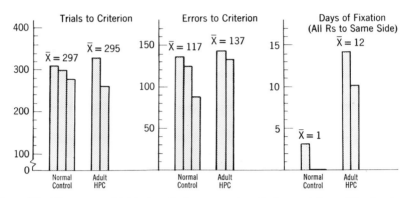

Figure 29. Histograms of trials to criterion, errors to criterion, and days of fixation to one side during the reversal phase of the brightness discrimination task (experiment 1).

1. For some problems, destruction of the hippocampus in the neonate produces less behavioral debilitation than that at later ages. This was found in our DRL operant tasks and in the insignificant impairment on the discrimination-reversal task. On the other hand, hippocampal destruction induced at any age

EXPERIMENT 2: REVERSAL

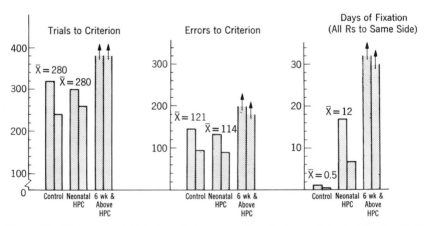

Figure 30. Histograms of trials to criterion, errors to criterion, and days of fixation to one side during the reversal phase of the brightness discrimination task (experiment 2). Bars with arrows represent animals which have not yet completed training.

produces a considerable impairment on the passive avoidance task. Thus the degree of impairment shown by animals with lesions made early in life is specific to the task.

2. For some behavioral problems destruction of the hippocampus at 6 weeks of age produces effects similar to those produced by neonatal damage (DRL performance), but on other tasks the lesion at 6 weeks of age produces as much impairment as an adult lesion (discrimination-reversal task, runway extinction, passive avoidance).

3. Hippocampal damage at 3 months produces the same degree of behavioral impairment as lesions made in adults in all tasks studied.

4. Even though neonatal damage to the hippocampus may lessen, or even eliminate, a particular deficit usually caused by lesions in adulthood, this does not imply that the behavior of the neonatally damaged animal is like that of normal animals. For example, the response latencies of animals leaving the start box in our runway, when the response was no longer rewarded, were longer in the neonatally damaged animals than those of control animals.

Although these conclusions must be regarded as tentative until all of our animals have completed the studies now underway and undergone histological evaluation, we believe them to be reasonable estimates of the complete picture that is emerging from our studies over the past three years. The view they provide may have to be somewhat altered in detail in the future, but we anticipate the perspective will be stable.

The results of destruction of the hippocampus, "sequelae," as we have called them, are many and mysterious. Yet they are somewhat less beyond our grasp than before—at least we hope so. The sequelae should not be evaluated as some unitary trend toward "normal behavior" but rather as a trend toward a different pattern of behavior than that produced by lesions made later in life. Of course, this observation fits in well without observations of the anatomical sequelae of the infant lesion as well. The histological evaluation of the animals shows a highly unusual anatomical configuration tending toward less destruction than would be anticipated on the basis of surgery, but by no means resembling a normal pattern of organization.

At one time we used the words "compensable" and "noncompensable" to describe problems on which the behavior of the neonatally damaged animal either does or does not show amelioration of the normal adult-lesion deficit. However, these terms tend to oversimplify the behavioral modifications produced by the lesions, since it is our impression that infant damage does produce its effects. These effects, however, are often different from those produced by adult lesions. Nevertheless, this categorization of the problems is of interest because there is evidence that tasks which are compensable in terms of early brain lesions are also ones in which animals with adult lesions can show improvements given changes in training conditions or the administration of drugs. This would imply that the compensable tasks tap abilities mediated by multiple systems or in which different strategies can be used.

The programs of research which must be accomplished in the future are twofold. We must first define more clearly the nature of the debilities produced in the noncompensable problems like the passive avoidance task in which lesions at any age produce deficiencies in behavior. At the same time we must define the nature of *compensable* tasks in order to understand the improved performance of early lesioned animals. The mere existence of the noncompensable problems indicates that neonatal damage always produces some irremedial effects. The problem now is to better understand their nature.

REFERENCES

1. Altman, J., and Das, G. D. Autoradiographic and histological evidence of postnatal hippocampal neurogenesis in rats. *J. comp. Neur.,* **124**:319-336, 1965.
2. Angevine, J. B., Jr. Time of neuron origin in the hippocampal region, an autoradiographic study in the mouse. *Exp. Neurol.,* Suppl. 2:1-70, 1965.
3. Doty, R. W. Functional significance of the topographic aspects of the retino-cortical projection. In R. Jung (Ed.), *The Visual System: Neurophysiology and Psychophysics.* Berlin: Springer-Verlag, 1960.
4. Douglas, R. J. The hippocampus and behavior. *Psych. Bull.,* **67**:416-442, 1967.
5. Jarrard, L. E., and Isaacson, R. L. Runway response perseveration in hippocampectomized rat: determined by extinction variables. *Nature,* **207**:109-110, 1965.
6. Jarrard, L. E. Isaacson, R. L., and Wickelgren, W. O. Effects of hippocampal ablation and intertrial interval on runway acquisition and extinction. *J. comp. physiol. Psychol.,* **57**:442-444, 1964.
7. Schmaltz, L. W., and Isaacson, R. L. The effects of preliminary training conditions upon DRL performance in the hippocampectomized rat. *Physiol. Behav.,* **1**:175-182, 1966.

EFFECTS OF INDUCTION AGE AND SIZE OF FRONTAL LOBE LESIONS ON LEARNING IN RHESUS MONKEYS

H. F. HARLOW, A. J. BLOMQUIST, C. I. THOMPSON,
K. A. SCHILTZ, and M. K. HARLOW

University of Wisconsin

In the mammalian brain, particularly the primate brain, there exist two extensive areas commonly described as cortical association areas. One of these, the frontal association cortex, anatomically defined as the frontal granular cortex, lies rostral to the arcuate sulcus, as illustrated in Figure 1, and its locus and extent can be described with precision. The second of these areas, the posterior association cortex, is more difficult to describe and localize. Insofar as visual associations are concerned, this area apparently includes the lateroventral

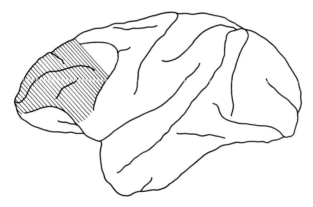

Figure 1. Frontal associative cortex (hatched area) of rhesus monkey brain.

This research was supported by U.S.P.H.S. grants MH-11894 and FR-0167 from the National Institutes of Health to the University of Wisconsin Primate Laboratory and Regional Primate Research Center, respectively. The authors are indebted to Drs. W. I. Welker, K. Akert, J. Johnson, and M. Murray for sharing in performing surgery on animals included in this study. They also express their gratitude to Miss Inge Siggelkow for handling and preparing the histological material.

79

portions of the posterior temporal cortex, the lunate sulcus, and some in-adequately defined cortical surface both rostral and dorsal to the sulcus luna-tus, as shown in Figure 2.

Damage to or destruction of the frontal association cortex in representative Old World monkeys, including rhesus, baboons, and mangabeys, impairs or destroys the ability of these animals to solve delayed response problems, as was first demonstrated by Jacobsen (9) and later confirmed by many other investigators, including Harlow and Settlage (8), Harlow, Davis, Settlage, and Meyer (7), and Pribram, Mishkin, Rosvold, and Kaplan (13). At the same time these extensive or even massive frontal lesions have been reported to leave other complex, visually guided learning, such as discrimination learning sets, relatively unaffected.

A radically different syndrome results from extensive lesions in the posterior associative cortex in catarrhine monkeys. Such lesions are reported to have little or no effect on the animal's delayed response capabilities (7) but com-monly result in serious impairment of discrimination learning and learning set performance. This differential loss associated with frontal and posterior le-sions of the neocortex has been described appropriately as a double dissocia-tion of symptoms: delayed response learning is greatly impaired by frontal lesions and little affected by posterior lesions, whereas discrimination learning and learning set formation are greatly impaired by posterior lesions and only slightly affected by frontal lesions.

There are limited data indicating that both the locus and size or mass of lesions produced in either the frontal or the posterior cortical association areas influence the severity of the specific learning deficits associated with damage to these areas. Jacobsen found that bilateral destruction of the sulcus principalis

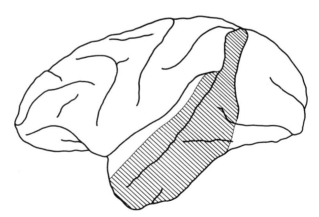

Figure 2. Visual association cortex (hatched area) of rhesus monkey brain.

and the surrounding dorsal and ventral cortical surface of the prefrontal lobes produced severe delayed response deficit, and Harlow and Settlage (8) reported that the severity and permanence of the impairment in delayed response performance by rhesus monkeys was influenced by the total mass of frontal tissue removed by frontal lobectomies.

Similar results relating to lesion size and locus have been reported following damage to the posterior association areas. Warren and Harlow (17), Chow (3), and Meyer (12) reported considerable sparing and partial recovery in some subjects on discrimination learning and/or learning set formation after bilateral extirpation of the temporal lobes in rhesus macaques. Some sparing was found by Raisler and Harlow (14) on discrimination learning set formation on multi-dimensional object problems and color problems after large bilateral temporal lesions combined with extirpation of the lunate sulcus and the surrounding parietal and occipital cortical surfaces, but no such sparing occurred for pattern discrimination problems.

There is now a considerable body of data indicating that behavioral loss is not only a function of location and size of lesion, but it is also a function of the age at which the lesion is induced. Recovery of coordinated movements following severe bilateral lesions in the cortical motor areas of monkeys was shown by Kennard (10, 11) to be a function of the monkey's age at the time of operation. Similarly, Akert, Orth, Harlow, and Schiltz (1) reported that the severity of delayed response impairment in rhesus monkeys subjected to bilateral lesions of the frontal associative cortex was a function of the animal's age at the time of lesion. Little or no loss occurred in monkeys operated on at 5 days of age as contrasted with severe loss occurring in animals lesioned at 24 months of age. Raisler and Harlow (14) reported age effects in rhesus monkeys undergoing two-stage posterior cortical lesions at 100 and 130 days, 340 and 370 days, and 870 and 900 days. Comparable age effects of cortical lesions have also been found in several other species of mammals tested on a variety of learning tasks, including maze learning (15) and pattern discrimination (16) in rats and roughness discrimination in cats (2).

The interpretation of the effects of age-dated cortical lesions on various learning tasks is complicated by the fact that the efficiency of performance of monkeys, and presumably of all mammals, on learning tasks is related to chronological age as well as cortical integrity. Aside from the human literature there exists only one study on the maturation of learning in primates (4), and this study is admittedly incomplete. Harlow showed that on a variety of learning tasks presented under standardized conditions there is a minimal age at which a kind of problem can be solved by rhesus monkeys after extensive training and an age at which the problem can be solved with maximal efficiency. Unfortunately the data are still lacking on the age at which maximal learning facility on the more complex problems is achieved by rhesus monkeys under standardized WGTA test conditions.

The maturation of ability to solve a discrimination problem employing a pair of stereometric objects, the first learning task presented to infant monkeys in the maturation study, is plotted in Figure 3. Results indicate that rhesus monkeys can solve the problem at 60 days of age in the WGTA, and they attain maximal efficiency by 120 to 150 days of age. The 60-day-old infants were the youngest group in the maturation study. Using slightly different methods, Zimmermann (18) was successful in training 20- to 30-day-old rhesus infants to discriminate between two objects, but these monkeys required more than six times as many trials as the 60-day group in the maturation study. Thus the ability to learn a visual discrimination problem is present in the first month of life and improves over the first four to five months in the rhesus monkey. If the difficulty of the discrimination were increased, for example, by using patterns instead of stereometric objects, the minimal age and age for maximal efficiency would probably increase.

The improvement with age in monkeys' ability to solve 5-second delayed responses is shown in Figure 4. By 120 to 150 days of age, most infant monkeys solve the problem in a reasonable training period, and they show striking

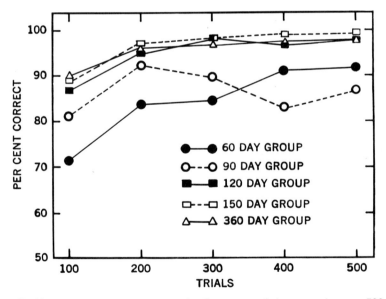

Figure 3. Mean percent correct responses for five groups of rhesus monkeys on 500 trials of a single discrimination problem. The group number designates the age of the group at the start of training.

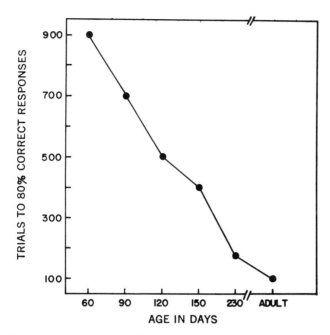

Figure 4. Mean number of trials on 5-second delayed responses for six groups of rhesus monkeys to achieve 80% correct responses. Age in days designates age of the group at the start of training.

improvement up to 8 months of age, when they are still inferior to rhesus adults. The changes between the first year and adulthood have not yet been determined.

Learning set solution employing multidimensional objects, particularly as measured by trial 2 performance, lies beyond the grasp of the rhesus monkey during the first five months of life. At about the sixth month the infant begins to make some small progress on this problem, and improvement is progressive from this time onward, as illustrated in Figure 5. Adult level of efficiency is not approached until the monkey is at least 18 months of age and probably older.

When the present neurological researches were designed in 1958, we assumed that learning set capability in the rhesus began to asymptote by 15 months of age, and this age was chosen to test learning set performance in our control group. We are now convinced that this conclusion was in error and are beginning testing of a second control group of subjects 2 years old at the start of training.

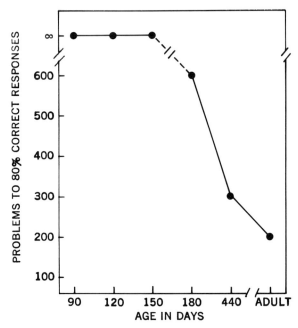

Figure 5. Mean number of 6-trial discrimination problems for six groups of rhesus monkeys to achieve 80% correct responses on trial 2. Age in days designates age of the group at the start of training.

The primary purpose of the present research was to determine the effect of lesions produced in the frontal association cortex of rhesus monkeys at ages ranging from 5 days to 2 years. Delayed response was the measure chosen as most likely to reveal lesion age effects. The strategy was to compare the effects of lesions produced before ability to solve delayed response has matured, at ages when the ability is increasing in normal animals and at an age when ability is presumably complete or nearly complete. Discrimination learning and learning set were chosen as measures likely to show maximal sparing even though learning set is a more difficult learning task than delayed response, as indicated by ages at which infants can begin to master the two problems and rate of improvement with increasing age.

A preliminary experiment conducted by Akert, Orth, Harlow, and Schiltz (1) subjected two 5-day-old monkeys to bilateral ablation of the dorsolateral prefrontal areas and compared their learning performance on a battery of tests with that of the ten normal monkeys of equal test age from the maturation study.

As can be seen in Figure 6, there were no significant differences in 5-second delayed response between the operated and normal monkeys, a finding contrary to that of most previous researches, which had imposed frontal lesions on mature rhesus macaques. When the same subjects were tested on 40-second delayed responses, the 5-day operated monkeys were originally inferior to the controls, but no differences existed by the end of a previously established standardized test series (see Figure 7). These same subjects were also tested on discrimination problems, string tests, and learning set. No learning differences between the operated and control monkeys were expected or obtained on these tests.

A second experiment conducted by Harlow, Akert, and Schiltz (6) compared the performance of two additional operated groups with that of the 5-day and control groups. Two animals were assigned to each of the two new groups, with frontal lesions produced in one group at 150 days of age and in the other group at 24 months. In spite of the limited number of subjects, the results appeared to be quite definitive, as shown in Figure 8. The acquisition curves for the 5-second delayed response trials are very similar for the control monkeys and those operated on days 5 and 150. These data stand in sharp contrast to

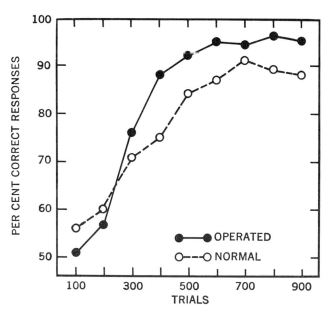

Figure 6. Mean percent correct responses of two rhesus monkeys topectomized at 5 days of age and ten normal monkeys tested at the same age on 5-second delayed response.

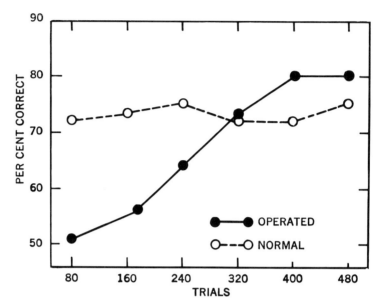

Figure 7. Mean percent correct responses of two rhesus monkeys topectomized at 5 days of age and ten normal animals of the same test age on 40-second delayed response.

those obtained from the monkeys lesioned at 24 months, which showed essentially no learning during the course of 900 delayed response trials. The only surprising finding was the efficient performance of the 150-day operates since these monkeys were lesioned at an age when delayed response capability had partially, even though not completely, matured. When the same groups of subjects were tested on 40-second delayed responses, the 150-day operated animals again gave the same picture as the 5-day operates (see Figure 9), with performance inferior to that of normal monkeys during the early training trials but equivalent during the last half of the series. The animals were also tested on discrimination learning, discrimination learning set, oddity learning set, and the Hamilton search test, and differences obtained between groups reflect differential ages at the time of testing.

Encouraged by the results of these preliminary researches we initiated and are continuing a major series of studies designed to measure the effects of frontal lesions induced in rhesus monkeys at various ages on a comprehensive battery of standardized tests. At the present time 16 normal monkeys are assigned to the study. Ten animals form a control group for early lesioned experimental subjects, and six will serve as a control group for 24-month-old experimental

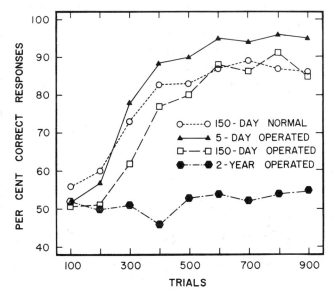

Figure 8. Mean percent correct responses of four groups of rhesus monkeys on 5-second delayed response. Age designation for operated groups is age at time of topectomy. Normal group and 5-day operated group began testing at 150 days of age; other groups were tested after recovery from surgery.

subjects. Groups of operated monkeys have been and are being formed to vary both age at time of operation and size of lesion. To date 24 monkeys have had bilateral frontal topectomies at ages varying from 5 days to 24 months, and 12 subjects have undergone bilateral frontal lobectomies at ages ranging from 2 months to 24 months. Twelve additional monkeys will be added at 48 months of age, divided into a lobectomy group, a topectomy group, and a control group.

METHOD

Subjects

The subjects whose data are reported in this paper are 44 rhesus monkeys, all but 4 of which were born in the laboratory. The exceptions were 2 Ss in the T-12 group and 2 Ss in the T-24 group, which had been captured in the wild and brought to the laboratory 1 to 4 months before their surgery. Their ages were estimated from dentition and weight. The remaining animals had varied rearing conditions and social experience, but none had participated in any learning or perception experiment with the exception of 4 animals in the 24-month

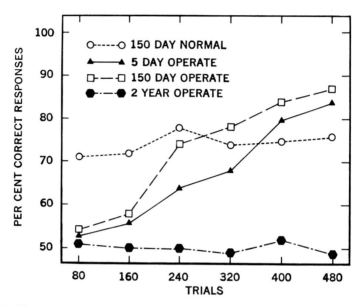

Figure 9. Mean percent correct responses of four groups of rhesus monkeys on 40-second delayed response. Age designation for operated groups is age at time of topectomy. Normal group and 5-day operated group began testing at 255 days of age.

topectomy group, which had completed 300 discrimination learning set problems 1 year prior to surgery. These Ss are not included in the results on learning set.

Animals were assigned to groups as shown in Table 1. Twelve animals were operated at either 50 or 150 days of age, when presumably the maturation of delayed response performance was incomplete, and 22 monkeys were operated at either 12, 18, or 24 months, at a time when they should have been performing at or near adult levels on delayed response tasks. On all tests the performances of the operated monkeys were compared with those of a new group of 10 normal controls since the particular adaptation and test procedures differed in certain respects from those used on infants in the maturation study (4). The two 5-day topectomized animals of the earlier report have been eliminated for the same reason.

Frontal operations were performed by aspiration and were completed in a single stage using standard operative procedures. Topectomies were designed to remove the frontal granular cortex in Brodmann's areas 9 and 10. The planned lesion is shown in Figure 10. The lobectomies included, in addition to Brodmann's areas 9 and 10, areas 11 and 12. All fiber systems underlying these four areas were ablated, but care was taken to spare the caudate nucleus. The planned lesion is shown in Figure 11.

TABLE 1. GROUP COMPOSITION

Group	Age at Surgery	Number of Monkeys
Controls	—	10
Topectomy		
T-5	5 months (150 days)	4
T-12	12 months	6
T-18	18 months	6
T-24	24 months	6
Lobectomy		
L-2	2 months (50 days)	4
L-5	5 months (150 days)	4
L-24	24 months	4

General Procedure

The test program is summarized in Table 2. All testing was conducted 5 days per week in a standard WGTA with appropriate accessories including test trays and stimulus objects. Raisins, small pieces of grape and apple, and, to some extent for older Ss, shelled corn were used as rewards, with individual Ss receiving the particular incentives to which they responded most consistently. Typically, older animals received at least two kinds of food during a session to maintain interest, but with infants grapes were used predominantly because the Ss preferred them.

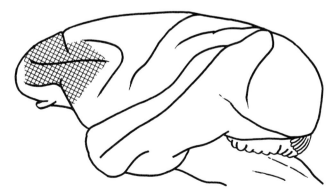

Figure 10. Area of intended lesion (cross-hatched) for bilateral frontal topectomy subjects.

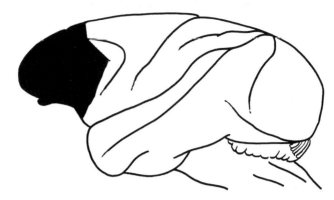

Figure 11. Area of intended lesion (black) for bilateral frontal lobectomy subjects.

Anatomical Method

Of the 34 operated Ss, 17 have thus far been sacrificed and their brains sectioned and studied microscopically. Two to four Ss are represented for each operate group except L-24, none of which has been sacrificed to date.

Standard diagrams were drawn from stained coronal sections of a normal, celloidin-embedded rhesus monkey brain. The sections of the normal brain were

TABLE 2. TRAINING SCHEDULE

Test Condition	Stimulus Objects	Daily Schedule	Number of Days
Adaptation-1	1 Object	25 Trials	See text
Adaptation-2	1 Object	25 Trials	See text
Adaptation-3	20 Objects	1 Object, 25 trials	20
Object discrimination	20 Pairs of objects	1 Problem, 25 trials	20
Delayed response (0, 5 seconds)	2 Red triangles	10 Trials at each delay	90
Delayed response (5, 10, 20, 40 seconds)	2 Red triangles	8 Trials at each delay	60
Object discrimination learning set	600 Pairs of objects	4 Problems, 6 trials each	150

cut at 30 microns and diagrams were drawn from sections 50, 100, 200, 300, 400, 500, 600, and 700.

The experimental animals were administered a lethal dose of pentobarbital sodium, which was followed by intracardial perfusion of 0.9% saline solution and then 10% formalin. The brain was removed and embedded in celloidin. Sections 30 microns thick were cut in a coronal plane and stained, and every tenth section was mounted on a slide. The experimental brain was then examined microscopically and the lesion drawn on a diagram.

Figures 12 to 17 show the location and size of the lesions for one representative animal from each of the six operated groups thus far sacrificed. It can be seen that the lesions include the areas intended and do not infringe upon adjacent structures.

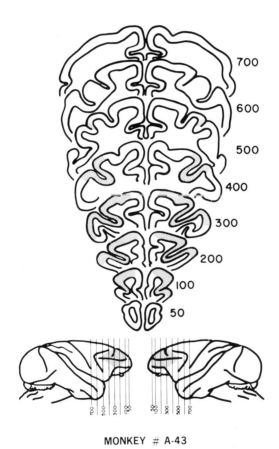

MONKEY # A-43

Figure 12. Lesion reconstruction of a representative 5-month topectomized monkey.

Behavioral Procedure

Adaptation

Procedures were changed midway in the study. One of the 10 control *S*s and 18 of the 34 operated *S*s were adapted by techniques in use in the Primate

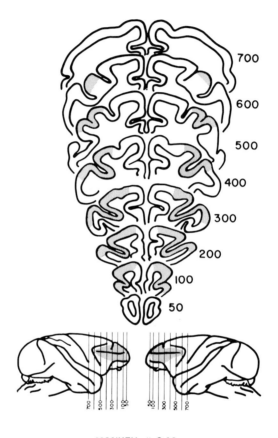

MONKEY # C-19

Figure 13. Lesion reconstruction of a representative 12-month topectomized monkey.

Laboratory for many years. The other *S*s were given modified procedures designed as an improved transition to the discrimination learning situation.

Adaptation 1. Both procedures began by training the animal 25 trials per day in the WGTA to obtain food from the centered well of a one-hole stimulus tray. When *S* responded within 5 seconds, a criterion reached in 1 to 3 days, an unpainted wood cube was placed behind the baited well and, on succeeding trials, moved forward until it covered the well. This procedure was continued until *S* responded by displacing the block consistently within 5 seconds for 1 week. The total time required varied from 3 to 6 weeks and completed adaptation for the first 19 *S*s.

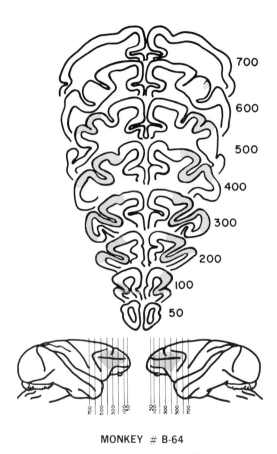

MONKEY # B-64

Figure 14. Lesion reconstruction of a representative 18-month topectomized monkey.

Adaptation 2. The remaining *S*s transferred to a two-hole board when they had reached the criteria outlined in adaptation 1. The unpainted cube shifted from left to right in a random order. Training continued until 30 test days (6 weeks) before their scheduled formal testing, at which time they began adaptation 3.

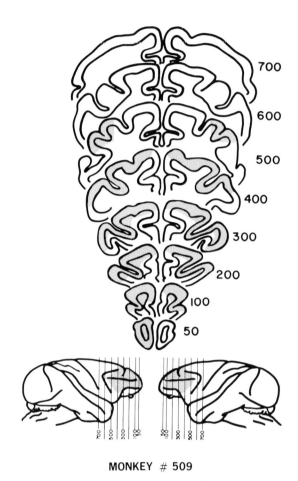

MONKEY # 509

Figure 15. Lesion reconstruction of a representative 24-month topectomized monkey.

Adaptation 3. This stage introduced a new object each day for 20 days, 25 trials with each object. The position of the object shifted in a randomized sequence. The monkeys were required to displace each object for 25 trials, and, as anticipated, some of the subjects balked when confronted by some of the new stimulus objects. The 10 extra test days had been provided to allow for repeating objects on which the Ss balked. If no balking occurred, objects were repeated for the remaining 10 days to keep the animals working until their formal testing date.

Most experimental Ss started adaptation training before surgery. Six operated animals (2 L-5's and 4 L-2's) had all their adaptation after surgery. Four

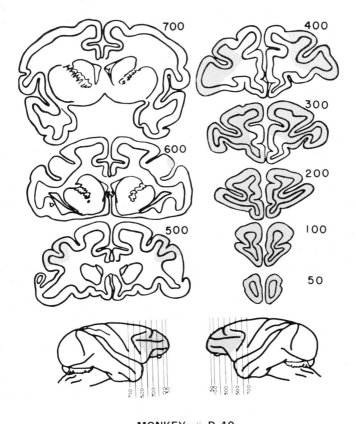

MONKEY # D-40

Figure 16. Lesion reconstruction of a representative 2-month lobectomized monkey.

operated animals (T-18's) had 3 days of preoperative adaptation, and the remaining Ss received 1 week or more of training before surgery. All animals had at least 3 and usually 6 or more weeks of postsurgical adaptation.

Object Discrimination

With the completion of adaptation training the animal was introduced to two-object discrimination problems in which one object was arbitrarily correct and covered a baited well and one object was incorrect and covered an empty well. The rewarded position shifted from left to right in random order balancing the two positions over 2-day periods with the additional condition that the same position would not be correct more than three trials in succession. A noncorrection procedure was used for this test and all subsequent tests. Twenty problems were presented, one per day, each utilizing a different pair of three-dimensional objects differing in multiple characteristics. There were 25 trials on each problem, a total of 500 trials for the 20 problems. A somewhat different procedure was used on two 150-day and two 24-month topectomized Ss tested in an earlier study (6), and these four Ss are consequently omitted in the present analysis of discrimination learning.

Delayed Response

Following object discrimination learning, Ss began training on the 0-second and 5-second delayed response task.* On each trial E gained S's attention before baiting one of the two food wells, then covered the wells with two identical red triangles. In the 0-second condition E immediately pushed the tray forward to within reach of S. Approximately 2 seconds elapsed between the time the food dropped into the well and the time the tray was within S's reach. The 5-second condition imposed a delay of that duration between the covering of the wells and the pushing forward of the tray. In all, approximately 7 seconds elapsed between the baiting and the time the tray was within S's reach. Both the position of the reward and the length of the delay were determined randomly with the restriction that a given position or delay would not occur more than three times in succession. Ten trials for each delay were presented daily, 5 days per week for 18 weeks, a total of 1800 trials.

Following the 90 days of 0- and 5-second delay problems, training began on a battery of delayed response problems with delay intervals of 5, 10, 20, and 40 seconds. The red triangles used for the earlier problems were again used, and 8 trials were presented for each interval each day, 5 days per week, for 12 weeks. Position of reward and length of delay were randomized with the restrictions that the position rewarded or the delay interval would not occur more than three times in succession. There were 1920 trials totally, 480 for each test interval.

Object Discrimination Learning Set

After the conclusion of delayed response testing, training began on a series of 600 object discrimination problems each 6 trials long. Problems were presented at the rate of 4 per day, 5 days per week, for 30 weeks. A new pair of objects differing in multiple dimensions was used for each problem.

---

*Four Ss, two T-5's and two T-24's, deviated somewhat in sequencing of tests. The T-5's had discrimination learning and training on the 0- and 5-second delayed response problems concurrently. The T-24's had three weeks of 0-second and 5-second delayed response training before they began concurrent training on discrimination learning. These two T-24's also had discrimination reversal problems concurrently with the battery of 5-, 10-, 20-, and 40-second delay problems. All four started discrimination learning set problems before their delayed response training was completed.

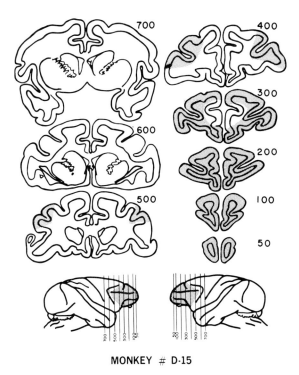

MONKEY # D-15

Figure 17. Lesion reconstruction of a representative 5-month lobectomized monkey.

Table 3 summarizes the numbers of Ss in the various groups and their mean ages at the start of each of the test batteries. The L-24 group has not completed learning set testing, and so the group N is 0 for the test. Four T-24 Ss had had previous training in learning set, and these Ss have therefore been omitted from the learning set results, reducing T-24 group to 2. Because of differences in procedure, as previously described, the T-5 and T-24 groups are both reduced by 2 on object discrimination learning. In general, the young subjects began testing as close to 200 days of age as possible logistically, and the 12-, 18-, and 24-month-old operated groups began testing as close as possible to 60 days after surgery.

Because age at the time of surgery varied widely among groups, it was not feasible to hold age of testing constant. The control group and the L-2, T-5, and L-5 groups were tested throughout at comparable ages. The T-24 and L-24 groups were similar to each other in age at testing. The remaining two topectomy groups were of intermediate age. A second control group starting testing at 2 years of age is planned for comparison with the T-18, T-24, and L-24 groups, but their testing has not yet begun.

TABLE 3.   NUMBER OF Ss AND MEAN AGE AT START OF TESTING
FOR EACH BATTERY OF TESTS

Group	Object Discrimination Task		0-5 Second Delay		5-10-20 and 40 Second Delay		Learning Set	
	Number of Ss	$\bar{x}$ Age (days)	Number of Ss	$\bar{x}$ Age (days)	Number of Ss	$\bar{x}$ Age (days)	Number of Ss	$\bar{x}$ Age (days)
T-5	2	200	4	206	4	346	4	409
T-12	6	423	6	465	6	611	6	728
T-18	6	630	6	660	6	790	4	878
T-24	4	790	6	802	6	945	2	920
L2	4	200	4	232	4	358	4	447
L-5	4	202	4	232	4	370	4	462
L-24	4	831	4	859	4	992	0	
Control	10	205	10	234	10	362	10	448

A difficulty exists because of the confounding of test age and operative age
for the groups as they now exist. It is probably minor for the delayed response
testing inasmuch as rhesus monkeys show large gains on the 5-second task up to
about 8 months of age and, although not fully mature in ability, they then approach adult ability. All groups were approaching or past this age when delayed
response testing began, and they were about 1 year of age or beyond when the
longer delay intervals were presented. Data from the earlier maturation study
had demonstrated that by 1 year of age animals perform near the adult level
on delays of 40 seconds under our standardized conditions.

Age is somewhat more crucial for the initial object discrimination problems
and object discrimination learning set problems. Although ability to learn a
single discrimination has matured before 200 days, when the youngest groups
started training, learning set ability has just begun to manifest itself and it continues to increase for at least a year longer and possibly several years. The
initial discrimination training was not designed as a learning set series but, rather,
was intended both to adapt the animals to a daily work program in which only
correct responses were rewarded and to add reliability to the method by controlling for chance stimulus preferences that may lead one S to errorless performance on a single problem and another S to persisting errors. Unfortunately,
the series was amenable to learning set training as well, thus handicapping the
younger animals. When the formal learning set problems began, the youngest
Ss were 409 to 462 days old and at some disadvantage as compared with the

T-12, T-18, T-24, and L-24 groups, all of which were 2 years of age at the start of the series.

RESULTS

General Behavior

In their behavior in their home cages and in the testing apparatus the operated animals and the control animals were indistinguishable. Although some experimental subjects were pacers, somersaulted, or circled in their cages or in the WGTA, especially during delayed response tests, an equal proportion of control subjects showed the same characteristics. One T-18 monkey showed more somersaulting than any other animal during testing, but it seemed not to handicap his performance. On delayed response he led the T-18 group. Convulsions were noted in only two Ss, both members of the L-5 group, and the seizures were observed only in the first 3 days after surgery.

Delayed Response

The results of primary concern are the effects of lesions of the frontal granular cortex on delayed response performance, and so we shall present these data first. They are probably the findings least confounded by age of testing.

Short Delayed Response: 0- and 5-Second Intervals

The data for the eight groups of monkeys were subjected to an unweighted means analysis of variance for unequal numbers of subjects. Within each of the eight groups the two delay periods (0 and 5 seconds) were analyzed as a repeated measure as were the nine successive blocks of 100 trials. The results for each group are plotted by interval and trial block in Figures 18 and 19. The 12-month and 18-month topectomy groups are presented separately (Figure 18) because they have no lobectomy or control group counterparts. The data for the eight groups collapsed across trials are presented in Table 4.

The three main effects and the interactions groups x delay, delay x trials, and groups x delay x trials were significant at the .025 level or beyond. Further statistical analysis using Fisher's LSD procedure and adopting the .05 level of significance disclosed no significant differences between the control group and the three groups operated at 2 or 5 months of age, but there were significant differences between both the control group and the L-2 group on the one hand and each of the four groups in which lesions had been produced at 12, 18, or 24 months. Combining the three early lesion groups and comparing their means using the individual degree of freedom technique with the combined T-12, T-18, T-24, and L-24 group means yielded a difference significant at the .01 level. Thus from a statistical point of view there is no evidence of delayed response

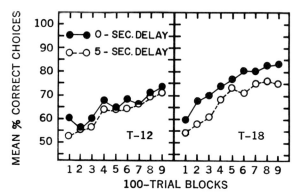

Figure 18. Mean percent correct responses of the T-12 and T-18 groups on 0- and 5-second delayed response presented in randomized order.

loss from lesions produced in the first half-year of life but demonstrable loss from lesions produced at 12 months or later.

The overall means for the control group and seven operate groups in Table 4 are ordered appropriately for our theses except for the mean of T-12, which performed less well than any other group, although it had the smaller of the two lesions and was younger than three of the groups at the time of operation. Two

TABLE 4. PERCENT CORRECT RESPONSES OF
NORMAL AND OPERATED GROUPS ON 0-, 5-, AND
0-SECOND AND 5-SECOND DELAYS COMBINED

Group	0 Second	Delay 5 Seconds	0 + 5 Seconds
Control	92	91	92
L-2	93	91	92
T-5	89	86	87
L-5	79	72	76
T-18	75	68	72
T-24	72	71	72
L-24	82	58	70
T-12	65	63	64

of the six subjects within this group showed only moderate loss, but four performed extremely ineptly.

Figure 19 presents curves of those groups that afford control of age and lesion size. We have plotted the learning curves for the normal subjects and the 50-day lobectomies and for the 5- and 24-month topectomized and lobectomized monkeys. These suggest that there may be a trend toward both age effects and size of lesion effects, even though there were no significant differences in means between T-5 and T-24, L-5 and L-24, T-5 and L-5, and T-24 and L-24. The performance of the L-24 group deserves special consideration. This group's attainment on the 0-second delays was well above expectation. At the same time they were by far the least efficient group on 5-second delayed response, a performance totally in keeping with predicted results for age of induction and lesion size. The T-24s, on the other hand, performed at a deficient but equal level on delays of both lengths. The explanation is not apparent to the authors.

Had we run all the groups only 300 trials, a more extensive test series than is usually employed, we would have obtained data much more in agreement with those generally reported for macaques subjected to frontal lobe lesions. The L-2 and normal monkeys would have performed nearly identically, but the T-5 group

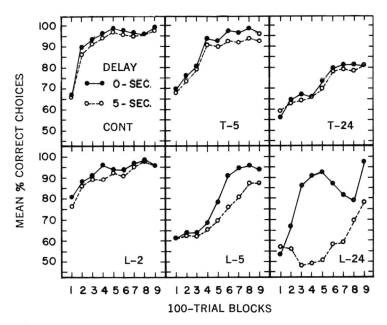

Figure 19. Mean percent correct responses of six groups of monkeys on 0- and 5-second delayed response presented in randomized order.

would have shown a modest deficit, the L-5 and T-24 animals a severe deficit, and the performance of the L-24 subjects would have been reported as approaching total failure. The phenomenon that recurs so frequently in our testing of the operated animals is that they start more slowly than normal animals, showing a positively accelerated curve early in testing and a negatively accelerated curve subsequently. This result suggests a deficit that is partially or totally compensated for by practice, the alternative being related to age and lesion size. The results at the end of extensive training are in better accord with the data reported on frontal lesion effects for other species than are the data on monkeys given abbreviated series of delayed response problems. Delayed response performance is not eliminated in chimpanzees by radical frontal lesions, and no defect on delayed response tests or on tests that are probably homologous has been discovered in human beings with frontal lobe lesions.

Multiple Delayed Responses: 5-, 10-, 20-, and 40-Second Intervals

The data obtained on the multiple delayed response series are very similar in overall trends to those previously reported from our laboratory (6). Since within obvious limits this is a replication as well as an extension of the previous work, the results give us basic confidence in our general conclusions regardless of significance levels for individual comparisons. Figure 20 presents the data for the eight groups collapsed across delay periods.

An unweighted means analysis was performed on the data, and within each of the eight groups the four delay periods (5, 10, 20, and 40 seconds) were treated

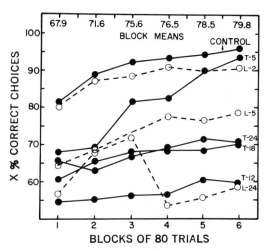

Figure 20. Mean percent correct responses of the eight groups on intermixed 5-, 10-, 20-, and 40-second delayed response. All intervals are combined.

as a repeated measure. The 480 trials run on each delay interval were broken down into six blocks of 80 trials each, and these six trial-blocks were treated as a repeated measure.

The main effects of groups and trials and their interaction were statistically significant at the .005 level. Delay interval did not interact significantly with any factor, but performance significantly decreased for all groups as delay became longer ($p < .001$).

The overall means for the eight groups are similar in trend to the means for the short delays. Again, the control group was unsurpassed by any operated group and, except for T-12, the groups show a decreasing efficiency of performance as lesion size and age at lesion increase. These data are presented in Table 5.

TABLE 5. PERCENT CORRECT RESPONSES OF
NORMAL AND OPERATED GROUPS ON THE
MULTIPLE DELAYED RESPONSE TESTS, 5, 10,
20, AND 40 SECONDS COMBINED

Group	Percent Correct
Control	91
L-2	88
T-5	81
L-5	73
L-18	68
T-24	67
L-24	61
T-12	57

There are essentially no differences between the control and the L-2 subjects, slight deficit for the T-5 group, greater deficit for the L-5 monkeys, and relatively severe deficits for the T-24 and L-24 animals. Once more the performances of the T-18 and T-24 subjects are almost identical; there is no evidence of increasing deficit developing following lesions at 24 months as compared with 18 months. The T-12 group is the least efficient group on multiple delay, as it was on the short delays. We have no explanation other than the probability that we have a chance selection of innately poor performers. This interpretation is strengthened by the finding that the T-12s also did badly on discrimination learning and learning set, tasks we expect to have less relationship to frontal lesions than does delayed response.

Analyses using Fisher's LSD procedure tested for differences between groups on multiple delayed response. The control monkeys were found not to differ significantly from the L-2 or T-5 groups, but they were significantly superior to the L-5 subjects and to all groups either topectomized or lobectomized at 12 months or older. Further, the young operated groups, those lesioned in the first 5 months of life, performed at a significantly superior level to those lesioned at or after 1 year of age.

In Figures 21 and 22 we have plotted the 10- and 40-second delayed response curves for groups in which age and lesion size are controlled—the normal subjects

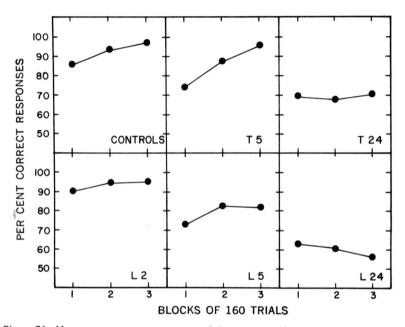

Figure 21. Mean percent correct responses of six groups on 10-second delayed response.

and the L-2, T-5, L-5, T-24, and L-24 animals. A striking phenomenon is the lack of evidence for any learning in the L-24 monkeys at either the 10- or 40-second delays. It was this group that had shown no evidence of learning on the 5-second delays during the first 500 trials of the short-delay tests and only slight improvement during the next 200 trials. Thus it would appear that in these 2-year-old lobectomized subjects we are approaching the point at which practice no longer ameliorates the effects of frontal lesions on delayed response performance as measured in our particular test situation.

Overall performance improved significantly across the six blocks of 80 test trials. Table 6 presents the percentages of correct responses on blocks 1 through 6 for all groups and delays combined. A significant groups x trials interaction, however, verifies the impression gained from Figure 20 that all groups were not improving at the same rate.

The length of the delay interval also had a clear-cut effect on delayed response proficiency, as shown in Figure 23, which plots percent correct responses as a function of length of delay. All these means are significantly different from

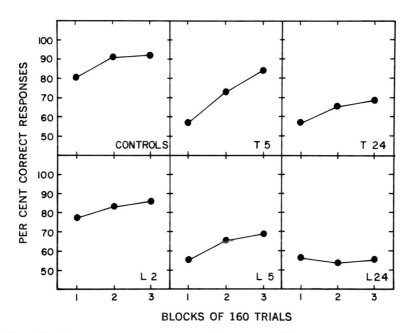

Figure 22. Mean percent correct responses of six groups on 40-second delayed response.

each other at the .01 level, and the actual drop in performance going from the 5- to the 40-second delay is over 13%, a finding that is in no way surprising.

Although the curves presented in Figures 20, 21, and 22 suggest that lobecto-mized monkeys performed at slightly lower levels than did their topectomized counterparts, this effect was not statistically significant as determined from a 2 x 2 unweighted means analysis of the T-5, T-24, L-5, and L-24 groups. In all these groups there were marked individual differences, with at least one subject performing relatively efficiently and at least one relatively inoffficiently. With

TABLE 6.   EFFECT OF PRACTICE ON DELAYED
RESPONSE PERFORMANCE FOR COMBINED
GROUPS AND DELAYS

Blocks	Percent Correct Responses
6	80
5	78
4	76
3	76
2	72
1	68

small numbers and high variability within groups, it would take very large differences in means to achieve statistical significance.

In spite of this we are convinced that bilateral lobectomy does produce more severe delayed response deficit than does bilateral topectomy. Although lesion size differences were not significant on the earlier 0- and 5-second delayed response measures or on the combined multiple delayed response performances (see Figure 24), the differences are consistent and in the same direction. This phenomenon also held for the age-matched groups in all four delay intervals considered separately in the multiple delayed response study. In absolute scores the performance of the T-5 monkeys, except at the very long delay intervals, was very similar to the performance of the L-2 group, the performance of the T-24 group resembled that of the L-5 group, and the performance of the L-24's was

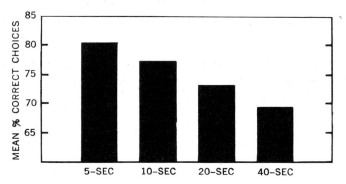

Figure 23.   Mean percent correct responses of the eight groups combined on delays of varying lengths in the multiple delayed response series.

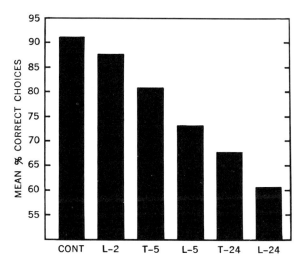

Figure 24. Mean percent correct responses of six groups on multiple delayed response, all intervals combined.

in a class by itself. Nonetheless, the data can be interpreted, probably properly, as indicating that the primary delayed response decrement is produced by destruction of the lateral surface of the prefrontal lobe and excision of the sulcus principalis (topectomy) and that bilateral lobectomy produces only a relatively small additional deficit.

We were initially surprised when our first two monkeys after bilateral topectomy at 5 days of age apparently showed no delayed response decrement whatsoever. We were far more surprised when four monkeys after bilateral frontal lobectomy at 50 days showed no loss, since the frontal operation was far more drastic and the time of operation somewhat longer delayed. Such data present strong presumptive evidence of vicarious functioning of other cortical areas which remain plastic in terms of their response capabilities for a period of time. The localization of such areas is, of course, a matter of complete conjecture; possibly the critical areas are in the posterior associative cortex or even in the limbic system, which do maintain afferent and efferent connections with the frontal granular cortex.

Object Discrimination Learning

The data for the eight groups of monkeys were subjected to an unweighted means analysis of variance for unequal numbers of subjects. The 20 problems were broken down into four blocks of 5 problems each, and problem blocks were analyzed as a repeated measure. Only trials 2 through 25 of each problem were included in the analysis because trial 1 is the informing trial and unsolvable by any principle.

The mean percentages of correct responses for the eight groups based on 480 trials per animal are given in Table 7. The groups have been arranged in descending order of performance. For the six groups affording control of age at time of testing, Figure 25 presents these data in graphic form.

TABLE 7.   MEAN PERCENT CORRECT RESPONSES
ON TWENTY LEARNING PROBLEMS

Group	Mean
T-24	91
T-18	86
Control	83
L-2	83
L-24	81
T-5	80
T-12	76
L-5	73

The overall group differences are significant ($p < .001$). Tests using Fisher's LSD procedure and adopting the .05 level of significance indicate that the T-24 group made more correct responses than all the other groups except T-18, and the T-18 group in turn was superior to the T-12 and L-5 groups. The ordering of the groups suggests both age and lesion-size effects and both were analyzed by further statistical tests.

To provide information about the effect of age and lesion size, Scheffe's method for multiple comparisons was applied to the four groups controlled for both variables. The T-24 and L-24 groups were combined and compared with the combined T-5 and L-5 groups to indicate age effects. The resulting difference is significant at the .005 level. Similarly, the T-24 and T-5 groups were combined and compared with the L-24 and L-5 groups combined. This difference is significant at the .025 level. Figure 26 plots the results of these tests. It is thus clear that animals operated at 2 years of age performed at a higher level than animals operated at 5 months, and animals receiving the small lesion performed at a higher level than their lobectomized counterparts.

The significance of the age effect is, we believe, a consequence of the age at testing and not the age at time of operation. Indeed, the data are in a direction

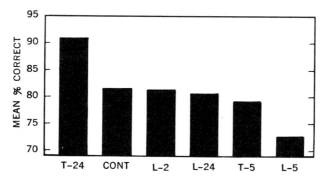

Figure 25.   Mean percent correct responses of six groups on 20 object discrimination problems.

opposite that which would be predicted if age at operation were the crucial factor. Inasmuch as we had not anticipated a significant deficit on this task as a result of either operation, we had not predicted the lesion-size effect or, of course, an effect related to age at operation. Moreover, we had not expected a large age-at-testing effect.

The discrimination test age results can only be explained by acceptance of the object discrimination series of 20 problems of 25 trials each as a learning set problem slightly simplified as compared with the conventional learning set series

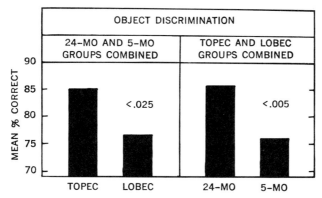

Figure 26.   Mean percent correct responses of combinations of 5-month and 24-month operated groups. Size of lesion is tested in graph on the left, and age at time of surgery is tested in graph on the right.

using problems of 6 or fewer trials. The data from the maturation study which indicated adult ability in discrimination performance at 4 to 5 months of age for rhesus infants are based on 500 trials with a single pair of objects. Similarly, the learning set data from the maturation study show no progress on learning set until 6 months and continuing improvement at least until 18 months. The learning set problems in the maturation study were 6-trial problems, which probably make learning set learning more difficult than do the 25-trial problems of the current discrimination task. The T-5 and L-5 groups were 200 days of age at the start of testing on the discrimination task, and the T-24 and L-24 groups were 790 and 831 days old, respectively, when they started testing. That the age factor was important is suggested by the superiority of the T-24 group over the normal animals tested at 205 days of age.

More detailed analysis of the data supports the explanation that learning set formation was occurring in the discrimination series. One evidence is a significant improvement in performance across blocks of problems ($p < .001$). For the eight groups combined, the percentages of correct responses on the succeeding blocks of five problems were 74.5, 81.5, 84.2, and 84.2, respectively. The Newman-Keuls test indicates a significant difference in performance on block 1 versus blocks 2, 3, and 4 combined.

The Wilcoxin matched-pairs signed-ranks test was applied to further comparisons of the data for evidence of interproblem improvement. The 20 problems were divided into two blocks of 10 problems each. Data from the 20 monkeys tested on object discrimination at about 200 days of age were analyzed separately from the data of the 20 monkeys tested after 1 year of age. A further breakdown analyzed separately for performance on trials 2 to 25, trials 2 to 6, and trial 2. The rationale for the separate trials analysis is the evidence (5) that interproblem learning is first apparent in the later trials of a problem. Thus there should be interproblem improvement manifest earlier on the trials 2 to 25 data than on the trials 2 to 6 data and earlier on the trials 2 to 6 data than on the trial 2 results.

Table 8 shows the results of these tests. For trials 2 to 25 only, both age groups showed interproblem improvement indicated by higher incidences of correct responses on the second block of 10 problems ($p < .01$, 2-tailed test). The trials 2 to 6 data indicate interproblem improvement only for the monkeys tested at over 1 year of age ($p < .02$). Trial 2 performance, which is the most rigid criterion of learning set, reveals no interproblem improvement for either group of animals. Thus these data considered together indicate some interproblem improvement in the young group and more interproblem improvement in the older group. The difference in interproblem inprovement in the two age groups doubtless was a factor in the superior performance of the older animals on the battery of 20 problems.

TABLE 8. COMPARISONS BY WILCOXIN *T* TEST OF PERFORMANCES OF YOUNG AND OLD SUBJECTS ON FIRST TEN DISCRIMINATION PROBLEMS VERSUS SECOND TEN DISCRIMINATION PROBLEMS

Test Age	Trial(s)	Mean Percent Correct Day 1-10	Day 11-20	*T* Value	Number Untied Pairs	*P*
200 Days	2-25	77.0	81.8	19	19	.01
	2-6	62.1	65.5	66.5	19	ns
	2	55.5	51.5	57	17	ns
Over 1	2-25	79.0	82.3	18	20	.01
year	2-6	64.0	71.1	42.5	20	.02
	2	60.5	60.5	51.5	14	ns

Meaningful comparisons which controlled for age of testing were possible between the control, L-2, L-5, and T-5 groups, all of which began object discrimination testing at about 200 days of age. Within this subset, the performance of the control group was not exceeded by that of any of the operated groups. Three orthogonal comparisons were made, using the individual degree of freedom technique, including *(a)* the control group versus the three combined operated groups, *(b)* the two groups operated at 5 months versus the group operated at 50 days, and *(c)* Group T-5 versus Group L-5. None of these differences approached significance.

In an attempt to minimize the effects of interproblem improvement in the older monkeys, an additional analysis of variance was run comparing all eight groups on only the first five object discrimination problems. Overall group differences were significant, with the rank order of the groups, from highest to lowest, as follows: T-24, T-18, L-2, L-24, control, T-5, L-5, and T-12. Subsequent tests (Fisher's LSD) indicated that the performance of Group T-24 was significantly superior to that of all other groups except T-18 and L-2 ($p < .05$). No other differences were significant.

In the consideration of group differences the 12-month topectomized monkeys again merit attention. Of the eight groups, they rank seventh in performance on the battery of 20 problems and eighth on the first 5 problems. They were 425 days old when they started testing—220 days older than the control group and 225 days older than the young operated groups. Their age should have given them an enormous advantage over the four young groups on the discrimination task, but their performance fails to reflect their age advantage. As on the delayed-response tests, the T-12 animals again appear to be aberrant.

Finally, attention should be called to the performance of the 50-day lobectomy group, which was the same age as the normal group at time of testing. Their error scores on the discrimination task were almost identical to those of the controls on the 20 problems and insignificantly higher on the first 5 problems. Thus on the discrimination task as well as on delayed response these animals with large lesions sustained at an early age could not be differentiated from normal animals.

In summary, older monkeys performed better than young monkeys on the object discrimination series of 20 problems, owing in part at least to the fact that they were developing learning sets to a greater degree than the young animals. When the effects of learning set are reduced by analyzing the data for only the first 5 problems, the three oldest groups still do well. Frontal damage thus does not drastically impair their performance even though there is a significant difference for size of lesion in the groups also controlled for age. Nonetheless, the group sustaining a large lesion at 50 days of age shows no evidence of any deficit on the discrimination task.

Object Discrimination Learning Set

Again an unweighted means analysis of variance was used, this time for seven groups, for none of the monkeys lobectomized at 24 months has completed object discrimination learning set (ODLS) training. The 600 problems were divided into six blocks of 100 problems each, and the problem blocks were treated as a repeated measure. Correct choices on trial 2 were used as the dependent measure.*

Individual group acquisition curves are presented in Figure 27. The overall difference between groups is statistically significant ($p < .05$), as is the improvement across the 600 problems ($p < .001$). The interaction between groups and problems is not significant.

Overall group comparisons are facilitated by Figure 28, in which scores are collapsed across the 600 problems. The results of subsequent tests using Fisher's LSD procedure are summarized at the bottom of the figure.

Interpretation of these data is complicated by the fact that monkeys in the various groups began testing at ages which ranged from 409 to 920 days and improvement in interproblem learning ability is probably continuous throughout this age range. Monkeys operated at 12, 18, and 24 months of age began

---

*An additional analysis was run using correct choices on trials 2-6 as the dependent measure. Other than producing an expected increase for all groups in the percentage of correct responses, owing primarily to intraproblem improvement, this latter analysis yielded results very similar to those obtained from the trial 2 data. Since interproblem improvement on trial 2 provides the more stringent criterion for learning set development, the results obtained from analysis of trials 2-6 are not discussed further in this paper.

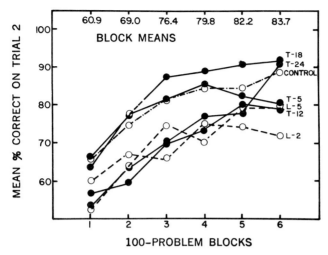

Figure 27. Mean percent correct trial 2 responses of seven groups of monkeys tested on 600 learning set problems of 6 trials each.

testing when they were 728, 878, and 920 days old, respectively, and their performance probably was somewhat superior to what it would have been had testing begun at ages more comparable to those of the remaining four groups.

Meaningful comparisons controlling for age were possible for the normal, T 5, L-5, and L-2 groups, all of which began testing when they were between 409 and 462 days of age. Within this subset, it is clear (see Figure 29) that none of the operated groups exceeded the control group in performance, although monkeys topectomized at 5 months made nearly as many correct choices as the control animals. Increasing the size of the frontal lesion, however, appears to have produced a marked decrease in learning set efficiency. The two lobectomized groups performed at a level at least 10% below that of the control and T-5 groups. Mann-Whitney U tests on subject scores collapsed across trials demonstrated no differences between the control and T-5 animals, but monkeys topectomized at 5 months of age performed significantly better on learning set than did the combined lobectomized groups ($p < .024$, 2-tailed).

Monkeys topectomized at 18 months made a slightly higher total of correct responses than the control group. This result would be expected if topectomy has little or no effect upon ODLS, for a test age difference of over 14 months would give a distinct advantage to the 18-month operates. The data for the T-24 group are based on only two animals. Although they made fewer correct responses than the control group, they were clearly capable of a high level of learning set performance. These animals were little above chance on the first

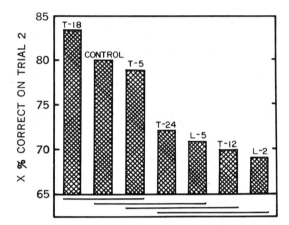

Figure 28. Mean percent correct trial 2 responses in 600 learning set problems for seven groups. Lines beneath bars indicate groups not differing significantly at the .05 level on Fisher LSD tests.

block of problems (see Figure 27) but improved steadily throughout testing, attaining a 91% level on the sixth problem block. This 72% gain over initial performance is larger than that of any other group. If topectomy was a factor in their early inferior performance, it seems not to be a factor in terminal performance, which shows no indication of asymptoting.

Other data suggest that the low early performance of the T-24 group is not a true picture of the effect of the operation on learning set learning. Four additional T-24 monkeys have completed 300 problems thus far. If their results are pooled with those of the first two animals, the mean percentage of correct trial 2 responses on problems 1 through 300 rises from 66 for the two to over 73 for the six animals. This compares favorably with the 74% correct responses attained by the control group and the 76% mean of the T-18 group on the first three problem blocks.

Once more, attention must be called to the 12-month topectomy group. As on all other tests, these animals performed far below expectation in terms of their age and lesion size. On the first three blocks of learning set problems, their trial 2 performance averaged 62%, the lowest attained by any of the seven groups and significantly lower than that of the T-18, control, T-24, or T-5 groups. On the 600 problems, their performance exceeded only that of L-2, from which they did not differ significantly. Their terminal performance was only 80%. As on all other tests, their inferiority is outstanding and an enigma.

In summary, the ODLS data suggest that removal of frontal granular neocortex produced little, if any, disturbance in ability to acquire an object

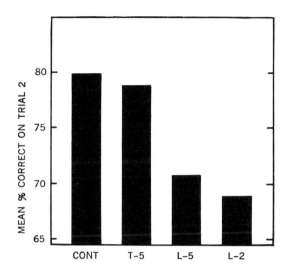

Figure 29. Mean percent correct trial 2 responses of the four groups tested on 600 object discrimination problems with age of testing constant.

discrimination learning set. The T-5 and control groups did not differ significantly, showing highly similar performances throughout the first 500 trials. Accurate assessment of any deficit in the older topectomy groups requires an additional normal group beginning learning set testing in the third year of life. It is clear, however, that the terminal performance of both T-18 and T-24 is high (91%) and, for T-24 at least, not leveling off. More severe frontal lobe damage produced a noticeable deficit on ODLS performance. The L-2 group was significantly inferior to the control and T-5 groups, and the L-5 group performed about 8% below the level of the control group. These losses in lobectomized animals far exceeded expectation both in terms of the locus of the lesion and the age at which the lesion was inflicted.

Summary and Discussion

The present research program was undertaken initially to test the hypothesis that bilateral frontal lobe lesions would result in increasing deficits in ability of rhesus macaques to perform delayed responses as the age at time of operation increases during the period in which the ability is developing. Excision of frontal tissue prior to the time delayed response ability is known to be manifest in normal rhesus infants should have minimal effects on subsequent ability to perform delayed response. Once the ability is manifest, age at time of operation should produce increasing deficits until the animals have attained full adult ability in this function. At this terminal developmental stage, frontal lesions should yield the extreme deficits typically reported for adult macaques with frontal lobectomies or topectomies. Moreover, it was anticipated that greater deficits would accrue from lobectomies than from topectomies at each age period. It was assumed on the basis of earlier research that frontal lesions would have little or no effect on visual discrimination learning or object discrimination learning set formation.

In general, the performance of the operated groups on delayed response in the present program shows a direct relationship between age at time of surgery and presence and degree of deficit. The initial two animals (6) were topectomized at 5 days of age and performed short delays without sign of deficit, even excelling the normal animals, although not significantly. On long delays they were initially inferior to normal agemates but subsequently improved to a level not significantly superior to that of normal subjects of the same age. Four groups of topectomized monkeys were included in the present report. Animals undergoing topectomy at 5 months, when delayed response ability is present, were slightly but not significantly inferior to control animals on all delays. Animals topectomized at 18 and 24 months, presumably at a time when ability is at least approaching the terminal level, performed at nearly identical levels significantly below those of normal animals. Aside from the failure to differentiate between animals topectomized at 18 months and animals topectomized at 24 months, which probably does not represent a deviation from the hypothesis in that terminal age for delayed response development is likely in the second year, there is an unexplained deviation of the 12-month topectomy group from the general pattern of decreasing performance with increasing age at time of surgery. The T-12 animals were usually the lowest performers of all topectomized groups.

Three groups of animals sustained bilateral frontal lobectomy at 2, 5, and 24 months of age, respectively. These larger lesions left the 2-month group essentially normal in the delayed-response tests, showing slight if any differences from the controls. Their performance was consistently superior to that of animals either topectomized or lobectomized at 5 months. The 5-month lobectomized monkeys, in turn, were consistently but not significantly inferior to 5-month topectomized animals. The 24-month lobectomized animals were generally, but not significantly, inferior to 24-month topectomized animals and the 5-month lobectomized animals. Unlike any other group, the L-24 animals showed essentially no improvement on long delays.

A repeated difficulty with delayed response testing in macaque monkeys is the high variability of performance from animal to animal even in normal groups, young or mature, and high variability on this test characterized all our operated groups and our normal group as well. We believe that the use of this test with small groups makes it difficult to establish statistically the presence of group differences. Thus we are as impressed with the continuity and consistency of declining performance means from young operates to older operates and from groups with small lesions to groups with large lesions as we are with the absence of expected statistically reliable differences between some of our groups.

The discrimination learning test and the object discrimination learning set test provided unexpected results. We had not anticipated that there would be performance deficits on discrimination tasks as a function of frontal lobe lesions. Even more surprising to us was the finding that size of lesion was related to performance. Still another unanticipated finding was the evidence for the beginning of learning set formation in the course of only 20 discrimination problems of 25 trials each. Although we knew that we lacked a control group for the older animals on the object discrimination learning set test, we had not expected that we would also need one for the initial discrimination learning test.

Among groups matched for age at testing, the youngest lobectomy group is indiscriminable from the normal animals in discrimination learning but significantly inferior to controls and to the T-5 group in object discrimination learning set. The L-5 subjects and T-5 subjects did not differ significantly from the controls or from each other on either task, although on learning set there was a systematic decrease in performance from T-5, to L-5, to L-2. The control and T-5 groups performed similarly on both kinds of discrimination tests. The effect of lesion size on object discrimination is apparent in the highly significant difference between the combined T-24 and T-5 groups and the combined L-24 and L-5 groups, and it is evident in the significant difference between the T-24 and L-24 groups. If age has a sparing effect on discrimination performance, it is not surprising that the T-5 and L-5 difference is not significant. It is unfortunate that there is no T-2 group to compare with the L-2 group on this and the learning set task. With the data lacking as yet on the learning set performance of some T-24 animals and all the L-24 animals, it is impossible to be definitive about lesion size effects at this age, but we now anticipate a significant difference between the lobectomy and topectomy groups.

When the data on lesion size are compared for the three tasks, we find that topectomy probably accounts for a large portion of delayed response deficit and lobectomy apparently adds little additional loss. For discrimination learning and object discrimination learning set tasks, on the other hand, apparently topectomy results in small losses, if any, whereas lobectomy produces significant deficit.

The most puzzling finding of all is the consistently inferior performance of the group topectomized at 12 months. Their test scores are inferior for both age and lesion size on all three types of test. Anatomical checks indicate that their lesions were of the intended size and locus. We tentatively conclude therefore that they represent an unfortunate selection of intellectually inferior monkeys, and we have started to study a new T-12 group to test this hypothesis. The only alternative explanation is that at 12 months of age rhesus monkeys are more susceptible to learning deficit on delayed response and discrimination-type tasks following a frontal lesion than are animals at least 7 months younger or 6 to 12 months older. At the present time this does not seem to be a reasonable or parsimonious hypothesis.

Especially noteworthy, we believe, is the amount of sparing of delayed response ability in all the groups except the 24-month lobectomy animals, which had very little success on delay intervals in excess of 5 seconds. Aside from them, no animals failed to show significant improvement with practice. Throughout the delayed response series there was overlap in performance between individual control animals and individual operated animals in all groups except L-24. Similarly, although the discrimination tests revealed deficits (especially in lobectomized animals), no group was unable to improve over the test series.

Although the data on learning set in the oldest animals are still incomplete, it is tempting to speculate about the significant loss in this function in the youngest lobectomy group. If the loss is a true one, it is the only significant deficit of any kind that we have found in animals operated on in the first 2 months of life. Inasmuch as discrimination learning set depends in part upon discrimination learning ability and discrimination ability is evident in rhesus monkeys in the first weeks of life, we may have tapped a loss in an already partly matured function. There is, however, the failure of the L-5 group to show an ever greater deficit: Theoretically, their deficit should have been greater than that of the L-2 group, and the L-24 group, on which we do not yet have data, should show a still greater loss in this ability. In the discrimination task L-5 did have a lower mean performance than L-2, and L-24, in spite of its age advantage, performed no better than any young group except L-5.

The data from this study thus far do not negate the general hypothesis that there is a relation between the amount of loss in an ability consequent to cortical damage and the stage of maturation of the ability that depends on the integrity of the cortical area subjected to lesion. Even if this hypothesis fails to hold after cumulation of data on multiple-type tasks and lesions in varied cortical loci, the hypothesis of a relation between age at the occurrence of cortical insult and subsequent performance level is not at stake. Accumulating evidence at the subhuman level indicates that a number of behaviors show sparing if the lesion occurs early in development, and it remains for future research to establish the limits of tenability. The development of other or more sensitive behavioral tests may uncover deficits not now apparent. Variations in location of lesions may reveal fixed specificities of function at variance with the findings for the frontal lesions in this investigation and for the sensory and motor area lesions of other experiments. Nonetheless, sufficient data are already available to force any theory of localization of function to give serious consideration to the age variable.

REFERENCES

1. Akert, A. K., Orth, O. S., Harlow, H. F., and Schiltz, K. A. Learned behavior of rhesus monkeys following neonatal bilateral prefrontal lobotomy. *Science,* **132**:1944-1945, 1960.
2. Benjamin, R. M., and Thompson R. F. Differential effects of cortical lesions in infant and adult cats on roughness discrimination. *Exp. Neurol.,* **1**:305-321, 1959.
3. Chow, K. L. Effects of temporal neocortical ablation on visual discrimination learning set in monkeys. *J. comp. physiol. Psychol.,* **47**: 194-198, 1954.
4. Harlow, H. F. The development of learning in the rhesus monkey. *Amer. Scientist,* **47**:459-479, 1959.
5. Harlow, H. F. Learning set and error factory theory. In S. Koch (Ed.), *Psychology: A Study of a Science.* New York: McGraw-Hill, 1959, pp. 492-537.
6. Harlow, H. F., Akert, A. K., and Schiltz, K. A. The effects of bilateral prefrontal lesions on learned behavior of neonatal, infant, and preadolescent monkeys. In J. M. Warren and K. Akert (Eds.), *The Frontal Granular Cortex and Behavior.* New York: McGraw-Hill, 1964, pp. 126-148.
7. Harlow, H. F., Davis, R. T., Settlage, P. H., and Meyer, D. R. Analysis of frontal and posterior association syndromes in brain-damaged monkeys. *J. comp. physiol. Psychol.,* **45**:419-429, 1952.
8. Harlow, H. F., and Settlage, P. Effect of extirpation of frontal areas upon learning performance of monkeys. *Res. Publ. Ass. nerv. ment. Dis.,* **27**:446-459, 1948.
9. Jacobsen, C. F. An experimental analysis of the frontal association areas in primates. *Neurol. Psychiat.,* **33**:558-569, 1935.
10. Kennard. M. A. Reorganization of motor function in the cerebral cortex of monkeys deprived of motor and premotor areas in infancy. *J. Neurophysiol.,* **1**:477-496, 1938.
11. Kennard, M. A. Relation of age to motor impairment in man and subhuman primates. *Arch. Neur. Psychiat.,* **44**:377-397, 1940.
12. Meyer, D. R. Some physiological determinants of sparing and loss following damage to the brain. In H. F. Harlow and C. N. Woolsey (Eds.),

Biological and Biochemical Bases of Behavior. Madison, Wis.: University of Wisconsin Press, 1958, pp. 173-192.

13.   Pribram, K. H., Mishkin, M., Rosvold, H. E., and Kaplan, S. J. Effects on delayed-response performance of lesions of dorsolateral and ventromedial frontal cortex of baboons. *J. comp. physiol. Psychol.,* **45**:565-575, 1952.

14.   Raisler, R. L., and Harlow, H. F. Learned behavior following lesions of posterior association cortex in infant, immature, and preadolescent monkeys. *J. comp. physiol. Psychol.,* **2**:167-174, 1965.

15.   Tsang, Y.-C. Maze learning in rats hemidecorticated in infancy. *J. comp. physiol. Psychol.,* **24**:221-248, 1937.

16.   Tsang, Y.-C. Visual sensitivity in rats deprived of visual cortex in infancy. *J. comp. physiol. Psychol.,* **24**:255-262, 1937.

17.   Warren, J. M., and Harlow, H. F. Learned discrimination performance by monkeys after prolonged postoperative recovery from large cortical lesions. *J. comp. physiol. Psychol.,* **45**:119-126, 1952.

18.   Zimmermann, R. R. Analysis of discrimination capacities in the infant rhesus. *J. comp. physiol. Psychol.,* **54**:1-10, 1961.

SPARING OF FUNCTION FOLLOWING LOCALIZED

BRAIN LESIONS IN NEONATAL MONKEYS

ARTHUR KLING and THOMAS J. TUCKER

Department of Psychiatry
University of Illinois College of Medicine

A growing body of experimental evidence suggests that localized ablations of the cerebral cortex in infancy frequently result in less severe functional deficits than similar lesions in juvenile or mature organisms.

Kennard (15, 16, 17), in a series of studies on the effects of ablations of motor and premotor cortex in monkeys at various ages, found that:

1. The earlier the lesion was made, the less deficit in motor function was observed. No deficit occurred in monkeys operated prior to 3 weeks of age. Lesions at later ages produced gradually increasing paresis but eventual recovery occurred if surgery was done prior to 7 months of age. Lesions sustained after 7 months resulted in spasticity, paucity of voluntary movement, difficulty in swallowing, and inability to right.

2. Extending the lesion to include either frontal association cortex or parietal lobe increased the disability, but lesions of these areas alone did not result in any motor deficit.

With respect to somatosensory function, Benjamin and Thompson (3) reported that lesions of somesthetic cortex, sustained prior to the seventh postnatal day in cats, produced little or no deficit in roughness discrimination, while Tsang (27) reported that maze learning was essentially unaffected with a variety of cortical lesions in the neonatal rat. Recently, Sharlock et al. (26) demonstrated that in the adult cat the capacity to discriminate differences in tonal duration was spared with lesions of auditory and related cortex sustained in infancy, and in an analogous visual task the present authors (28) found considerable sparing of a flux discrimination following lesions of areas VI and VII in cats operated from 4 days to 2 months of age. Deficits in discrimination learning began to appear in those cats operated at 2 months of age, and more marked deficiencies occurred in the adult-operates. Also in the visual modality, Doty (7) reported a later

---

This research was supported by a grant from the United States Public Health Service, National Institute of Child Health and Human Development, Number 02277 and in part by the Spastic Paralysis Research Foundation, Kiwannis International.

sparing of pattern vision with lesions of visual cortex in the neonatal kitten. However, infant-operates were as severely impaired as the adult-operates if the lesion was extended to include nonstriate areas as well.

In a series of studies in our laboratory (9, 18, 19, 20), amygdalectomy in the infant cat, rat, and monkey did not result in changes in affective, sexual, or oral behavior commonly seen after similar lesions in adults of these species. In the kitten there was some acceleration in the rate of motor development and a suggestion of precocious puberty in the female. At the time of puberty, some elements of the typical adult syndrome appeared but did not persist. In the monkeys sacrificed prior to puberty (between 2 and 3 years of age) no evidence of the syndrome had yet appeared. Objective testing of affective responses revealed only minor changes in the juveniles compared to major changes in adults with similar lesions.

In comparing the effects of hippocampal lesions in infant and adult cats, Isaacson (13) found a distinct difference between these groups in the resistance to extinction of a simple runway response. In addition, the deficit in discrimination reversal learning, found in the group of adult-operates, was greatly attenuated in the infant preparations. On the other hand, both infant- and adult-operates were similarly impaired in the learning of passive avoidance and DRL 20-second schedules of reinforcement. Isaacson suggested that these latter deficits were not compensated for by other neural systems, thus resulting in equally severe deficits in both groups.

Beach (2) reported that those cortical lesions found to be effective in radically altering maternal behavior in the primiparous rat were quite ineffective if made in the prepubertal animal.

With respect to cognitive function, Harlow et al. (10) reported a later sparing of delayed response performance after incomplete ablations of frontal granular cortex in two infant monkeys. In animals operated at later ages there appeared to be a corresponding reduction in the degree of this sparing. Prefrontal ablations at 8 months of age and beyond produced no delayed response sparing whatsoever. Raisler and Harlow (25) studied the effects of lesions of posterior association cortex in 130-, 370-, and 900-day-old monkeys. They found that the 900-day-old group was retarded on object discrimination, whereas the younger animals were not.

From this review it is quite apparent that the phenomenon of functional sparing after localized cerebral ablations in the neonatal animal has been demonstrated for some functions of the somatomotor system, somatosensory system, vision, audition, affective-sexual behavior, and some cognitive functions.

Although current empirical evidence continues to accrue in support of this phenomenon, there have been few attempts to conceptionalize the phenomenon in a theoretical framework. In the early forties Kennard was of the opinion that, with respect to motor function, the remaining cortex subserved voluntary motion, since extending the lesion either anteriorly or into sensory cortex resulted in a more profound deficit, and that, prior to 3 weeks of age (in the macaque), motor performance was largely subcortical, thus allowing for considerable alteration in cortical development and establishment of alternative functional pathways.

More recently, Isaacson, in studying hippocampal function, divided the effects into two main categories—those capacities which are spared as a function of age at time of injury, or "compensatory functions," and those which are lost regardless of what age the injury is sustained, or "noncompensatory." However, the basis for this differentiation is as yet unknown.

From our studies on the amygdala, it seems that while a definite lack of effect is seen with neonatally sustained lesions, the total behavior at any one point in development is a reflection of the maturational state of remaining brain. Accordingly, some aspects of the syndrome of amygdalectomy became prominent at the time of puberty, probably as a result of the action of gonadal hormones on subcortical structures. Kennard reported that with advancing age, a degree of spasticity developed in some animals previously free of the symptom. This raises the issue as to the course of the functional sparing with maturation. Most of the studies referred to test the animal at a relatively early age, when the brain is still not fully mature.

Another issue about which little is known concerns the role of subcortical structures in the mediation of functional sparing, since most of the previous work has been limited to the cerebral cortex.

The purpose of this paper will be to review a series of experiments directed toward a fuller understanding of those neural mechanisms involved in the sparing of cognitive and motor function after neonatal ablation of dorsolateral frontal granular cortex and related cortical and subcortical areas in the monkey.

I. DIFFERENTIAL EFFECTS OF LESIONS OF DORSOLATERAL FRONTAL GRANU-LAR CORTEX IN THE INFANT AND ADULT MONKEY

It has been well established that bilateral destruction of dorsolateral frontal granular cortex in the juvenile or mature monkey produces a syndrome characterized by an inability to perform the delayed response task beyond a 5-second delay (24) as well as marked hyperactivity and distractability (8, 22). Although the hyperactivity may diminish in time, the deficient performance on delayed response does not significantly improve with the lapse of time or prolonged training (14).

The purpose of this experiment was to examine the effects of extensive lesions of dorsolateral frontal granular cortex in neonatal monkeys on later delayed response performance, to observe for the development of hyperactivity or distractability and to compare the performance of maternally reared subjects with some who were maternally-deprived.

METHODS

All infant monkeys (M. mulatta and M. speciosa) used in this study were born in our laboratory of imported pregnant females or laboratory matings. Adult subjects were obtained from commercial sources. The infants were operated between the first and thirty-fourth postnatal day and adults at least 1 month after arrival. All subjects were anesthetized with pentobarbital sodium (¼-cc/lb body weight for adults; .20 cc for a 450 to 500 gm infant). Bilateral aspiration of prefrontal cortex was accomplished in one stage via a frontotemporal craniotomy.

After surgery, infant-operates were kept in an incubator for 24 hours, after which all but one were returned to their respective mothers for rearing. EF-102 was artificially reared in order to compare her performance with maternally reared counterparts. The maternally reared subjects were not separated from their mothers until the initiation of testing, between 5 and 10 months of age. Adult subjects began testing 2 months after operation. All monkeys were maintained on a reduced diet during formal testing. Following several days of habituation to the WGTA, training began on a red-green color discrimination problem, using apple slices as reward. For detailed procedure see (29). After criterion performance was established on the color problem, the monkeys began delayed response training. Initially, the subjects were trained on 0-second delay interval to a criterion of 90% correct responses in a daily 30-trial test session. Subsequent delay intervals tested were 5, 10, 20, and 40 seconds. Training was discontinued if the subject failed to achieve criterion performance within 500 trials on any of these delay intervals. After the maximum delayed response performance was determined, each subject was tested on a 5-second delayed alternation task. A total of 1020 trials in 17 sessions were given to each animal and the total percent of correct responses was computed.

Following completion of behavioral testing, the subjects were killed and the brains removed for histological verification of the lesion and retrograde thalamic degeneration.

RESULTS

Results are shown in Figure 1.

Lesions

On the basis of the surface reconstructions and thalamic degeneration, most infant and adult subjects sustained complete ablations of dorsolateral frontal granular cortex. There was an incomplete removal of tissue within the bend of the arcuate sulcus (left hemisphere) in EF-102, the right hemisphere of LF-201, and left hemisphere of LF-202. In general, the extent of destruction was somewhat greater in the infant-operates than in the adults. Orbitofrontal cortex was largely spared in all frontal-operates, as was the head of the caudate nucleus.

With the exception of LF-202 (left hemisphere), the principal sulcus was entirely ablated in all subjects. Retrograde thalamic degeneration was confined to the paralamellar and parvocellular divisions of the dorsomedial nucleus with some additional cell loss in the ventral anterior nucleus. No detectable differences were noted in the degree of nuclear cell loss or gliosis between the infant- and adult-operates.

General Observations

Recovery from surgery was uneventful. The infants were alert, vocalized, and accepted bottle feeding 6 to 12 hours after operation. They were readily accepted by their mothers and immediately oriented to the nipple and began to nurse. They displayed no evidence of hypotonia, lethargy, or hyperactivity. They showed appropriate responses to maternal communications, were gradually separated and weaned to solids. The nature of their play, aggressive behavior, and responses to humans was no different than that seen in unoperated, laboratory-reared animals.

The artificially reared monkey (EF-102) developed the syndrome characteristic of maternally deprived monkeys, as described by Harlow (11). This included excessive non-nutritive sucking and marked fear responses, such as curling up, hair-pulling, and rocking in a corner of the cage. All responses were exaggerated when she was placed in a novel situation or when disturbed by strangers. Remarkably, this behavior did not seriously interfere with her testing ability.

Somatic growth up to 1 year of age was within normal limits for all lesioned subjects.

The hyperactivity and circling behavior, characteristic of the adult-operated subjects, was not seen in any of the infant-operates up to 1 year of age. However, EF-104, who was kept for a more prolonged period (2 years), gradually developed both characteristics at approximately 1½ years of age, and they persisted until he was sacrificed (24 months).

FORMAL TESTING

No essential differences were found between the infant- and adult-operated groups on the color discrimination problem. All animals reached criterion performance within approximately the same number of trials. EF-104 maintained his performance after retesting 1 year later.

Test performance on 5-second delayed alternation also failed to reveal any significant differences between the early- and late-lesioned subjects. None of the subjects mastered this task.

However, a marked performance difference occurred on delayed response. Without exception, all infant-lesioned subjects were capable of criterion performance at longer delay intervals than were any of the adult-operates.

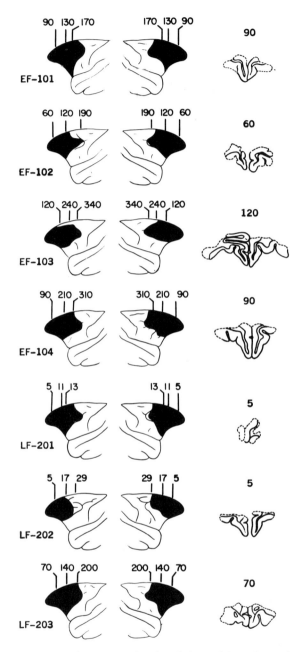

Figure 1. Reconstructions of dorsolateral surface lesions of frontal granular cortex (translated to a standardized view of the monkey brain) and selected coronal sections for four infant-lesioned and three adult-lesioned monkeys.

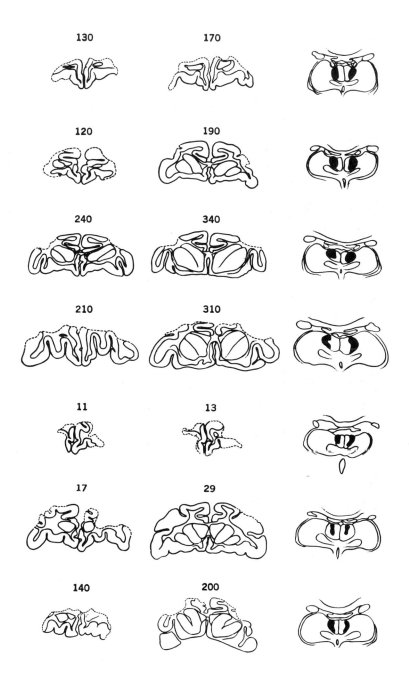

TABLE 1. TEST PERFORMANCES OF EARLY- AND LATE-LESIONED MONKEYS ON COLOR DISCRIMINATION, 5-SECOND DELAYED ALTERNATION AND DELAYED RESPONSE

| Groups | Age at Operation | Age at Initial Testing (months) | Color Discrimination (Trials to Criterion) | | 5-Second Delayed Alternation (Percent Correct Trials) | | Delayed Response | | | |
| | | | | | | | Maximum Delay (seconds) | | Trials to Attain Maximum Delay | |
			First Testing	Second Testing	First Testing	Second Testing	First Testing	Second Testing	First Testing	Second Testing
Early frontals:										
EF-101(M)[a]	2 days	5	150		42		40 (6 months)	40 (8 months)	320	130
EF-102(M)	1 day	5	60		37		10		750	
EF-103(M)	4 days	6	60		49		40		1526	
EF-104(S)	34 days	10	60	60 (22 months)[b]	55 (14 months)	62 (23 months)	40 (10 months)	5 (18 months)	960	180
Late frontals:										
LF-201(S)	3 years (approximate)	38	90		39		0		210	
LF-202(S)	3 years (approximate)	38	60		49		0		180	
LF-203(S)	3 years (approximate)	38	120		41		5		810	

[a](M) indicates M. mulatta; (S) indicates M. speciosa.
[b]Age at testing given in parentheses following the test score.

Three of the four infant-operates were capable of doing 40-second delays, whereas the longest delay achieved by any adult-lesioned subject was 5 seconds. Although normative data were not obtained in this experiment, the infant-lesioned frontals did as well as unoperated subjects of similar age, as reported by Harlow et al. (12).

Retesting of delayed response in EF-101 at 8 months revealed a somewhat better performance than at 6 months. However, retesting EF-104 at 18 months (9 month interval), indicated that performance efficiency had decreased to a level (5 seconds) characteristic of adult-operates.

II. COMBINED LESIONS OF DORSOLATERAL FRONTAL GRANULAR CORTEX AND CAUDATE NUCLEUS IN THE NEONATAL MONKEY

Since it had been shown that extensive bilateral frontal cortex ablations in the neonatal monkey failed to result in the characteristic delayed response and locomotor impairments (at least to 1 year of age), the present experiment was designed to extend the lesion into a related subcortical structure in an attempt to elucidate the neural mechanisms involved in the functional sparing.

In the adult monkey, bilateral lesions of the head of the caudate nucleus also produce delayed response impairments (1) and hyperactivity (6). Anatomical evidence of corticotopic connections between prefrontal cortex and caudate nucleus have been demonstrated (23).

In the adult a prefrontal-caudate system seems to be implicated in the disturbances seen after either cortical or caudate lesions alone; therefore we prepared a group of neonatal monkeys with the combined frontal-caudate lesion in order to determine whether the sparing soon after cortical lesions alone would still be present (21).

In addition to supporting certain cognitive functions, it has been suggested by Kennard and others that early somatomotor development and control is dependent on subcortical structures. Therefore subcortical injury in the neonatal period may be expected to produce deficits in motor function and reflex control not found after cortical injury.

This experiment was designed to compare the delayed response performance of monkeys sustaining combined lesions of frontal granular cortex and caudate nucleus with those sustaining only the cortical lesion and to compare the development of early reflexes, motor and affective behavior, growth and survival.

METHODS

Seven infant M. speciosa sustained bilateral ablations of dorsolateral frontal granular cortex and head of caudate nucleus between 2 and 56 days of age. The surgical procedure was similar to that described previously, except that after the cortical ablation had been accomplished, the underlying white matter was incised, exposing the dorsal surface of the caudate nucleus. Since the gross distinction between gray and white matter is poor in the infant, an attempt was made to visualize the anterior horn of the lateral ventricle and then to aspirate medially. It was difficult to visualize the extent of the lesion at the time of surgery, so that considerable variation in the size of the caudate lesion was to be expected.

After surgery, the infants were placed in an incubator. On the following day, some were returned to their respective mothers for rearing and some were kept for artificial rearing.

Formal testing began when they reached 7 months of age. As in the previous experiment, the testing sequence involved the acquisition of a red-green discrimination, delayed response from 0 to 40 seconds, followed by 5-second delayed alternation.

Observations of early reflexes, somatic growth, motor function, and affective behavior were made throughout their survival period. Surviving subjects were killed between 12 and 14 months of age and their brains removed for histological verification of the lesion and retrograde thalamic degeneration.

RESULTS

Results are shown in Figure 2.

Lesions

In general, the extent of the frontal granular cortex lesions were similar to those reported previously except for additional damage to orbital cortex (in some cases). In IFC-303 there was sparing of the caudal portion of the principal sulcus.

Varying degrees of injury to the caudate nucleus were present in the three subjects. In no case was there a complete lesion, the degree of caudate destruction varying between 10 and 50%. IFC-301 had the smallest lesion, which damaged only the rostral caudate in the left hemisphere. The remaining two had more extensive but asymmetric lesions. Damage to surrounding structures included the internal capsule, putamen, and claustrum.

Retrograde Degeneration

Retrograde degeneration was present in all three divisions of the dorsomedial nucleus. The degeneration in the magnocellular division was a consequence of the injury to orbitofrontal cortex. Additional thalamic degeneration was observed in the following nuclei: ventral anterior, anteroventral, anteromedial, ventrolateral and reticular.

Survival and Rearing

In contrast to the infants sustaining frontal cortex lesions alone, none of the seven frontal-caudate preparations were successfully reared by their mothers. Four, who were operated between 7 and 15 days of age, were returned to their

TABLE 2. REARING, SURVIVAL

Lesion	N	Survive 150 Days		Cause of Death		
		Mat.-Reared	Mat.-Deprived	Operative	Infection	Failure to Thrive
Frontal	8	5	1	1	1	0
Frontal–caudate nucleus	7	0	3	0	1	3
Frontal–P.T.O.	4	0	3	0	0	1 (premature)

mothers as in the previous experiment. Although they were readily accepted, they appeared to become weak, hypokinetic, and malnourished, were frequently found on the floor of the cage, did not appear to be nursing, and in time became emaciated and died. When it became apparent that these infants were not thriving, they were removed from the home cage and an attempt was made to treat them, but it was unsuccessful. Survival times were 1 week for two subjects and 6 weeks for the other two. At autopsy, one of these was found to have a hydrocephalus and meningo-encephalitis. No gross pathology was found in the other three. All three surviving subjects were artificially reared in incubators for at least 3 months. Two of the three (IFC-302, IFC-303) had poor sucking reflexes for the first two postoperative weeks. They had to be fed by trickling the milk into their mouths or allowing them to chew on the nipple. Vomiting was frequent and weight gain was very slow during the first 2 months. As sucking became stronger, food intake increased. Their first 2 months were marked by poor temperature control, requiring frequent application of external heat. In general, they were weak, hypotonic, and hypokinetic. A summary of the data on survival and rearing is presented in Table 2.

Somatomotor development was slow. IFC-302 had a flaccid paresis of the right leg until 2½-months of age. By 3 months, her gait appeared to be normal. IFC-303 had a profound weakness in both legs, which gradually improved to a normal status by the second postoperative month.

Clinical seizures affected all surviving frontal-caudate preparations during the first 2 postoperative weeks. IFC-301 was observed to have three focal seizures of the left arm and leg. IFC-302 had many focal seizures of the left arm, and IFC-303 had several generalized convulsions involving both the face and extremities. No clinical seizures were observed after the second postoperative week. Repeated scalp EEG's gave evidence of continued seizure activity and abnormal slowing in IFC-302 and IFC-303. IFC-301 had only slow frequency, low amplitude records.

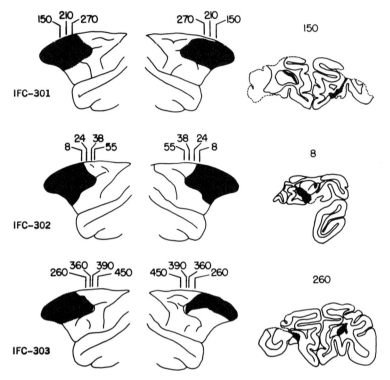

Figure 2.   Reconstructions of dorsolateral surface lesions and selected coronal sections for three infant-lesioned frontal-caudate monkeys.

As the frontal-caudates matured, they all developed excessive fear responses, withdrawal, bizarre posturing, and other behaviors characteristic of the maternally deprived monkey. Episodic bursts of self biting and screaming occurred frequently during the testing sessions, especially in response to error.

Unlike the frontal preparations, all of the combined-lesioned group displayed excessive hyperactivity, circling, and distractability, which became accentuated during testing.

Growth

The surviving frontal-caudates showed deficient growth until they were sacrificed. No accurate measure of caloric intake was made, but they appeared to eat as much as other monkeys and their food preferences were no different. Part of the explanation may lie in the fact that they suffered more gastrointestional and upper respiratory infections, which resulted in periodic anorexia.

From the ovarian weights of the two females in the group, there was an indication of retardation in gonadal development, whereas adrenal size was unaffected.

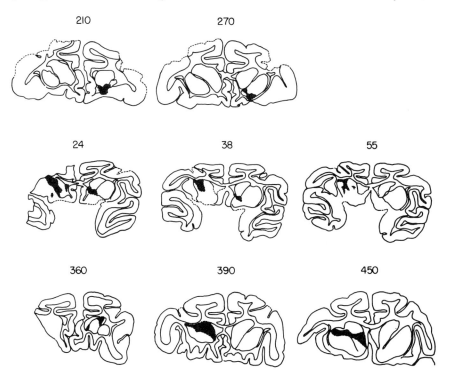

Results of Formal Testing

As shown in Table 3, the frontal-caudates performed no differently on color discrimination than did those with frontal lesions alone. They also did no better than chance on delayed alternation, again similar to the frontal preparations. However, with the exception of IFC-301, who had only minimal damage to the caudate, the other two frontal-caudates failed to achieve criterion performance beyond 0-second delay. The delayed response deficits of IFC-302 and IFC-303 are comparable to those of adult monkeys with frontal cortex lesions alone.

III. COMBINED FRONTAL AND POSTERIOR ASSOCIATION CORTEX LESIONS IN THE INFANT AND ADULT MONKEY

In experiment I it was found that lesions restricted to frontal association cortex, when produced in infancy, failed to result in delayed response and locomotor deficits, at least through 1 year of age. Extending the lesion to include the head of the caudate nucleus, as in experiment II, resulted in a failure

TABLE 3. TEST PERFORMANCES OF INFANT-LESIONED FRONTAL AND FRONTAL-CAUDATE MONKEYS ON COLOR DISCRIMINATION, DELAYED RESPONSE, AND DELAYED ALTERNATION

Groups	Age at Operation (days)	Age at Initial Testing (months)	Color Discrimination (Trials to Criterion)	Delayed Response Maximum Delay (seconds)	5-Second Delayed Alternation (Percent Correct in 1020 Trials)
Infant frontals[a]:					
IF-101(M)[b]	2	5	150	40(180)[c]	42(243)
IF-102(M)	1	5	60	10(152)	37(316)
IF-103(M)	4	6	60	40(193)	49(296)
IF-104(S)	34	10	60	40(313)	55(435)
Infant frontal-caudates:					
IFC-301(S)	56	7	65	40(215)	49(302)
IFC-302(S)	15	7	50	F[d](221)	52(240)
IFC-303(S)	2	7	60	0(215)	54(248)

[a]Data for infant frontal monkeys were reported previously (28).
[b](M) indicates M. mulatta, (S) indicates M. speciosa.
[c]Age (in days) at time of testing given in parentheses.
[d]F indicates failure to reach criterion within 500 training trials.

on delayed response, in addition to the hyperactivity and distractability characteristic of the adult preparation.

In the adult monkey, posterior association cortex (parieto-temporo-preoccipital) has been implicated in both delayed response performance (5) and sensory discriminations (4). Since it was found that a combined cortical-subcortical
lesion in the infant prevented the sparing that follows a frontal lesion alone, the
question arose as to whether, in the infant, extension into another *cortical* region
(i.e., a combined cortical-cortical lesion) related to the task in question would
have a similar effect, or whether, even after large lesions of cerebral cortex, there
can still be a sparing of the delayed response task.

METHODS

Three infant female monkeys (one M. mulatta and two M. speciosa) were
operated on in two stages: the left hemisphere from 3 to 6 days of age and the
right hemisphere from 13 to 17 days of age. Two separate bone flaps were turned
on each side; one temporoparietal flap for exposure of the posterior association
cortex and one anterior to the coronal suture for the frontal cortex lesion. The
operative procedure was essentially the same as described previously. One infant
(IO-503) was returned to its mother for rearing but was rejected and subsequently
reared artificially. The other two M. speciosa, which were born 1 week apart,
were reared together until the beginning of testing at 7 months of age when they
were placed in separate cages. Two adult preparations sustained lesions similar
to the infants.

In addition to color discrimination, delayed alternation, and delayed response,
all subjects were tested on object discrimination.

Object discrimination testing was conducted in the WGTA; this required that
the animal learn to consistently choose one of two simultaneously presented
three-dimensional objects which differed in size, shape, color, and texture. A
correct choice was rewarded with a small slice of apple. For example, one pair
of stimulus objects consisted of a small, circular, yellow metal ashtray paired
with a large, square, green slice of rubber sponge. If the monkey chose the ashtray in at least 27 trials out of a consecutive 30, training began on a new pair of
objects until they were also learned to a comparable level. A total of 10 such
object-pairs were presented to the animal, and the total number of trials required
to master all 10 problems were computed for each animal.

LESIONS

The ablations of frontal granular cortex were complete in all but one case. In
IO-502, the prefrontal region of the left hemisphere remained largely intact.

The parieto-temporo-preoccipital lesions, which were extensive on surface
reconstructions, did not always reach the depth of the sucli so that there was

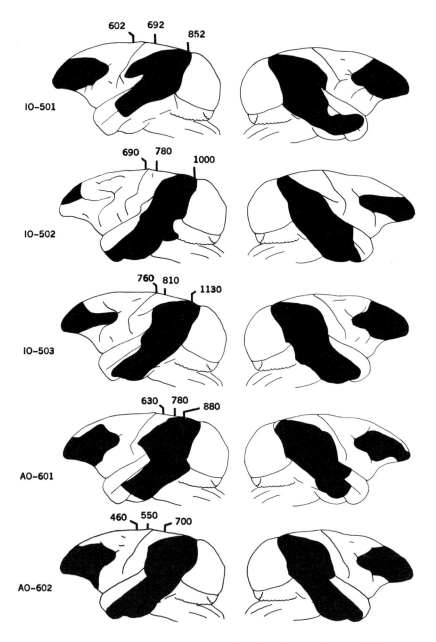

Figure 3. Reconstructions of dorsolateral prefrontal and posterior association cortex lesions for three infant-lesioned and two adult-lesioned monkeys.

136

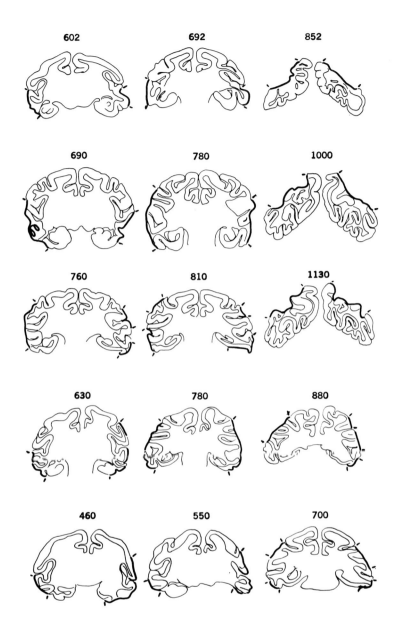

602   692   852

690   780   1000

760   810   1130

630   780   880

460   550   700

variation in the depth of the lesion between subjects. In all cases there was sparing of the most ventral portions of the temporal neocortex. It was intended not to injure the superior temporal gyrus, but, as seen in Figure 3, this occurred in cases AO-601 and IO-501. The more complete resections were achieved in the posterior parietal and preoccipital regions.

Thalamic degeneration in the dorsomedial nucleus was consistent with that observed in the previous experiments. It was expected that the pulvinar would show more extensive retrograde degeneration than was the case. We estimated that, in most cases, only 10 to 20% of the cells in the pulvinar showed any degeneration, mainly in the inferior and lateral portions. A more complete anatomical description will be presented in a forthcoming publication.

RESULTS

Recovery from surgery was uneventful in all three infant-operates. Although they displayed some general weakness and hypokinesia for several weeks, they could suck well and grasp, and they were alert and responsive to stimulation. Unlike the frontal-caudate infants, there was no evidence of seizures, paresis, or abnormalities in reflexes.

Two of the infants (IO-502 and IO-503) were reared in the same cage. They would spend most of the time clutching one another and displayed marked discomfort when separated for feeding and cleaning. Their usual position was a ventral-ventral embrace. IO-502, which was 1 week younger than IO-503, gradually became the more dependent of the two, clinging to IO-503 as an infant to its mother. During their first 3 months they both displayed non-nutritive sucking on the digits and ears of each other but not on their own body. They did not display the self-clutching, rocking, hair pulling, and withdrawal characteristic of their maternally deprived counterpart, IO-501. However, after they were separated at the initiation of testing (7 months), they both gradually developed self-clutching, withdrawal, bizarre posturing, excessive fear responses, and non-nutritive sucking. Both were fearful in the testing apparatus, with IO-502 exhibiting the more extreme behavior. When given the opportunity to be together, they would clutch one another as they had before separation. Both preparations displayed considerable emotionality until they were sacrificed.

As seen in Table 4, both infant and adult preparations mastered the color discrimination problem within the range of subjects with frontal and frontal-caudate lesions. However, on object discrimination the performance of infant-operates was far superior to that of the adults. Another major difference was seen in delayed response performances. The infant-operates mastered all delay intervals tested, whereas the two adult-operates failed to achieve 0-second delay. The results for the infant-operates are consistent with those of infants sustaining only the frontal ablation, indicating that extension of the frontal cortex lesion to posterior association cortex fails to alter the sparing of this cognitive task.

TABLE 4. TEST PERFORMANCE OF INFANT-LESIONED FRONTAL-PTO AND ADULT-LESIONED FRONTAL-PTO MONKEYS ON COLOR DISCRIMINATION, OBJECT DISCRIMINATION, DELAYED ALTERNATION, AND DELAYED RESPONSE

Group	Color Discrimination (Trials)	Object Discrimination (Trials)	Delayed Alternation (Percent Correct in 1020 Trials)		Delayed Response	
			5-Second Delay	3-Second Delay	Maximum Delay (second)	Trials
Early frontal-PTO:						
501	60 (6 months). 60	570	56 (5 months). 70		40 (5 months). 40	540(5 months).220
502	150	760	90 (1 month). 90		40	620
503	90	580	46 (1 month). 54	90[a]	40	1719
Late frontal-PTO:						
601	60	1860	46		F (2 months).. F	
602	60	1290	48	50	F ..(1 month).. F	

[a]Achieved Criterion (90 percent correct) on 10 sec. Delayed Alternation.

However, the infant frontal-operates performed no differently than did the adult-operates on the delayed alternation task, just as in the two previous experiments.

DISCUSSION

Maternal rearing was possible in those infants sustaining frontal cortex ablations and, from a previous study (20), after amygdalectomy. In both cases the infants showed no disturbances in sucking or grasping, thus allowing for adequate maternal stimulation. One of the three with the combined frontal-posterior cortex lesion was returned to the mother but was rejected and subsequently hand-reared. We did not attempt to return the other two for fear of losing any of the sparse laboratory births, but on the basis of their adequate sucking and grasping, we would guess that these preparations have the potential for being maternally reared. In contrast, none of the frontal-caudate preparations were successfully reared by their mothers. Their general lethargy, poor sucking and grasping ability, and possibly their early seizures led to inanition and death. On the basis of the size of the lesions for the three artificially reared subjects, we would say that the larger the subcortical damage, the greater the deficit in early reflexes and activity levels. However, even relatively small lesions of the basal ganglia were sufficient to cause severe debilitating effects and a high mortality in the neonates. These data provide evidence for the notion that the integrity of subcortical systems is more significant than an intact cerebral cortex for survival of the neonatal monkey. However, recent data from our laboratory on the effects of *very extensive* cortical lesions indicate that:

1. Neodecortication in the infant monkey results in severe paresis of the legs with only gross movements possible in the arms and hands, poor sucking ability, poor temperature control, general weakness, and, with advancing age, development of spasticity.

2. Sparing of only temporal lobe (bilaterally) does not result in as severe a deficit but, nonetheless, would make adequate maternal stimulation or survival highly unlikely.

In our experience, the only extensive neocortical lesion in which maternal rearing might be possible, is that which spares motor and premotor cortex. In three subjects sparing these areas, good motor strength, adequate sucking, and alertness may provide for sufficient somatomotor ability to make maternal rearing possible.

Although subcortical systems may be crucial to motor performance and other survival functions in the neonate, it will be recalled that in our two infant-operated subjects with striatal damage who showed postoperative paresis, there was a gradual recovery of motor function by 3 months of age. Since, by comparison, no recovery took place after neodecortication, only moderate recovery after sparing of temporal lobe and good residual somatomotor function with sparing of motor and premotor areas, we believe that a considerable representation of motor function does, in fact, exist in the neonatal cortex. Accordingly, it is probable that the motor recovery following early striatal injury occurs at a cortical level. The superb motor performance of our preparations sparing only motor and premotor cortex, in contrast to the temporal lobe or neodecorticate preparations, would suggest that there exists in the neonatal cortex a potential for function consistent with the classical hypothesis of functional localization.

An interesting finding related to motor development was the appearance of hyperactivity and circling at 5 to 6 months of age in the three frontal-caudate monkeys. This behavior is characteristic of adults with frontal cortex lesions alone or combined frontal-caudate injury. It was not observed in our neonatal preparations with frontal or combined frontal-posterior cortex lesions, at least up to the age of 16 months. However, one of our infant-lesioned frontal subjects (EF-104), who was kept longer than the others, began to display this classical behavior at approximately 18 months of age. This behavioral syndrome then appeared first in those infant-operates with the combined cortical-subcortical lesion and only much later in one subject with the prefrontal lesion alone. It would seem likely then that both cortical and basal ganglia structures participate in the suppression of the hyperactivity and circling, but the basal ganglia exert the greater suppressing effect early in development.

The sparing of delayed response performance in the frontal and combined frontal-posterior groups, at least through 1 year of age, is contrasted with the failure of the frontal-caudate preparations on this task. It would seem unlikely, in the light of the good performance by the frontal-posterior group, that the sparing of this task following a frontal lesion alone depends upon the integrity of remaining association cortex. Rather, it would seem that the integrity of basal ganglia is an important link in the preservation of the ability to perform this task in the neonatal-operate. Just as an intact striatum is initially important for optimal somatomotor development in the neonate, so it appears to be critical in early problem-solving, or cognitive, capacity as well. But unlike the eventual recovery in somatomotor function following early striatal injury, there is still no evidence for a later compensation for the initial cognitive loss.

An important finding, based on only one animal, concerns the fact that the initially superb delayed response performance of the frontal monkey may progressively deteriorate with continued maturation. Since the cognitive deficit appeared at about the same time as did the hyperactivity (18 months), it would not be unreasonable to suppose that the disturbance in some central mechanism responsible for the cognitive deficit is also directly or indirectly responsible for the disturbance in locomotor function. Accordingly, it may be argued that the capacity to suppress, or at least adequately regulate, motor activity is an important requirement for good performance on the delayed response problem. At least through 1 year of age, the infant-operates with restricted cortical lesions have this ability and are able to perform adequately. But, as Kennard found with motor cortex lesions, a progressive deterioration may occur with time.

The question of why some tasks can be compensated for and others cannot remains a major issue for discussion. It will be recalled that none of our infant-lesioned frontal or frontal-posterior subjects showed any sparing of the delayed alternation task, even though they were capable of performing quite efficiently on the delayed response problem. These results suggest that an analysis of task differences, with respect to the kinds of requirements for solution inherent in each, may provide a key to understanding the differential sparing obtained. From the literature review presented earlier, it would seem that the kinds of learning tasks showing the greatest degree and persistence of compensation are those that rely heavily on the presence of discriminable sensory cues for their solution. Similarly, it would seem that injury of primary sensory projection areas in the neonate is less likely to result in permanent functional losses than is injury sustained in either motor or "associational" cortex, as traditionally defined.

The inability of our infant-lesioned subjects to master the delayed alternation task may be due to the fact that the solution of such a problem depends primarily upon the establishment of a sequential order, or principle, of responding which is relatively independent of (i.e., not guided by) any extrinsic sensory aspect of the test situation per se. In addition to the absence of overt sensory discriminanda, the subject is required to perform an avoidance response on each trial, in the sense of avoiding the particular stimulus location (in a binary choice situation) that "paid off" in the immediately preceding trial. Clearly, these complications do not obtain in the case of the delayed response problem, where the sensory aspect of the baiting phase and the direct approach of the subject during the choice period are fundamental to the solution of the problem. Accordingly, the extent to which learning tasks are or are not spared subsequent to neonatal brain injury may largely depend upon the total number of elements required in task solution (i.e., task complexity) and the amount of exteroceptive sensory information necessary to that solution. In all fairness, it must be added that although this hypothesis may serve to delineate the *type* of behavioral task most likely to survive the effects of neonatal brain injury, we have, as yet, no substantial information on the nature of those neuronal events on which the functional sparing effect is based.

SUMMARY

1. Ablations of dorsolateral frontal granular cortex in the neonatal monkey do not result in impairment on the delayed response task through 1 year of age. Both infant- and adult-operates were equally impaired on delayed alternation. There is a suggestion that the sparing of delayed response may begin to diminish at approximately 18 to 24 months of age.

2. Combined lesions of frontal granular cortex and posterior association cortex in the neonatal monkey also result in sparing of delayed response as well as object discrimination through 1 year of age. Adult-operates were severely impaired on both tasks.

3. Combined lesions of frontal granular cortex and caudate nucleus resulted in failure on delayed response in both infant- and adult-operates. Unlike the two previous groups, this lesion resulted in poor survival, postoperative seizures, temporary paresis, retardation in growth, and appearance of hyperactivity at 5 to 6 months of age.

4. Maternal rearing was possible after frontal cortex ablations but not when combined with injury to the basal ganglia.

5. Somatomotor maturation may be related to performance efficiency on delayed response.

6. It is suggested that the degree and persistence of behavioral sparing, following neonatal brain injury, may be related to the factors of task complexity and sensory cue dominance.

ACKNOWLEDGEMENTS

We acknowledge the assistance of Mrs. Clara Preston in the preparation of the brain tissues, Miss Joan Gilmour in the rearing of the brain-operated infants, and the technical assistance of Mr. Clifford Sherry and Miss Renee Fox.

REFERENCES

1. Battig, K., Rosvold, H. E., and Mishkin, M. Comparison of the effects of frontal and caudate lesions on delayed response and alternation in monkeys. *J. comp. physiol. Psychol.,* **53**:400-404, 1960.

2. Beach, F. A. The neural basis of innate behavior: Relative effects of partial decortication in adulthood and infancy upon the maternal behavior of the primiparous rat. *J. Genet. Psychol.,* **53**:109-148, 1938.

3. Benjamin, R. M., and Thompson, R. F. Differential effects of cortical lesions in infant and adult cats on roughness discrimination. *Exper. Neurol.,* **1**:305-321, 1959.

4. Blum, J. S. Cortical organization in somesthesis: Effects of lesions in posterior associative cortex on somatosensory function in Macaca mulatta. *Comp. Psychol. Monogr.,* **20**:(3), 219-249, 1951.

5. Blum, J. S., Chow, K. L., and Pribram, K. H. A behavioral analysis of the organization of the parieto-temporo-preoccipital cortex. *J. comp. Neurol.,* **93**:53-100, 1950.

6. Davis, G. D. Caudate lesions and spontaneous locomotion in the monkey. *Neurology* (Minneap.), **8**:135-139, 1958.

7. Doty, R. W. Functional significance of the topographical aspects of the retino-cortical projection. In R. Jung and H. Kornhuber (Eds.). *The visual system: Neurophysiology and Psychophysics.* Berlin: Springer-Verlag, 1961.

8. French, G. M. Locomotor effects of regional ablations of frontal cortex in rhesus monkeys. *J. comp. physiol. Psychol.,* **52**:18-24, 1959.

9. Green, P. C., and Kling, A. Effect of amygdalectomy on affective behavior in juvenile and adult macaque monkeys. *Proc. Amer. Psychol. Assoc.,* New York, 1966.

10. Harlow, H. F., Akert, K., and Schiltz, K. A. The effects of bilateral prefrontal lesions on learned behavior of neonatal, infant and preadolescent monkeys. In J. M. Warren and K. Akert (Eds.), *The Frontal Granular Cortex and Behavior,* New York: McGraw-Hill, 1964, pp. 126-148.

11. Harlow, H. F., and Harlow, M. K. The affectional systems. In A. M. Schrier, H. F. Harlow, and F. Stollnitz (Eds.), *Behavior of Nonhuman Primates,* Vol. II, New York: Academic Press, 1965, pp. 287-334.

12. Harlow, H. F., Harlow, M. K., Rueping, R. R., and Mason, W. A. Performance of infant rhesus monkeys on discrimination learning, delayed response, and discrimination learning set. *J. comp. physiol. Psychol.,* **53**:113-121, 1960.

13. Isaacson, R. Cytoarchitectural and behavioral consequences of neonatal lesions in the cat. Informal presentation given before Michigan Association of Mathematical, Experimental, and Physiological Psychologists, Inglis House, April, 1966.

14. Jacobsen, C. F. Studies of cerebral function in primates I. The functions of the frontal association areas in monkeys. *Comp. Psychol. Monogr.,* **13**:(3), 1-60, 1936.

15. Kennard, M. A. Age and other factors in motor recovery from precentral lesions in monkeys. *Am. J. Physiol.,* **115**:138-146, 1936.

16. Kennard, M. A. Reorganization of motor function in cerebral cortex of monkeys deprived of motor and premotor areas in infancy. *J. Neurophysiol.,* **1**:477-496, 1938.

17. Kennard, M. A. Cortical reorganization of motor function: Studies on a series of monkeys of various ages from infancy to maturity. *Arch. Neurol. Psychiat.* No. 8, 227-240, 1942.

18. Kling, A. Amygdalectomy in the kitten. *Science,* **137**:429, 1962.

19. Kling, A. Behavioral and somatic development following lesions of the amygdala in the cat. *J. Psychiat. Res.,* **3**:263-273, 1965.

20. Kling, A. Ontogenetic and phylogenetic studies on the amygdaloid nuclei. *Psychosom. Med.,* **28**:155-161, 1966.

21. Kling, A., and Tucker, T. J. Effects of combined lesions of frontal granular cortex and caudate nucleus in the neonatal monkey. *Brain Res.,* 6:428-439, 1967.

22. Malmo, R. B. Interference factors in delayed response in monkeys after removal of frontal lobes. *J. Neurophysiol.,* 5:295-308, 1942.

23. Nauta, W. J. H. Some efferent connections of the prefrontal cortex in the monkey. In J. M. Warren and K. Akert (Eds.), *The Frontal Granular Cortex and Behavior,* New York:McGraw-Hill, 1964, pp. 397-409.

24. Orbach. J., and Fischer, G. J. Bilateral resections of frontal granular cortex. *Arch. Neurol.* (Chicago), 1:78-86, 1959.

25. Raisler, R. L., and Harlow, H. F. Learned behavior following lesions of posterior association cortex in infant, immature and preadolescent monkeys. *J. comp. physiol. Psychol.,* 60:167-174, 1965.

26. Sharlock, D. P., Tucker, T. J., and Strominger, N. L. Auditory discrimination by the cat after neonatal ablation of temporal cortex. *Science,* 141:1197-1198, 1963.

27. Tsang, Y.-C. Maze learning in rats hemidecorticated in infancy. *J. comp. Psychol.,* 24:221-253, 1937.

28. Tucker, T., and Kling, A. Differential effects of early vs. late brain damage on visual duration discrimination in cat. *Fed. Proc.,* 25(2), March-April 1966.

29. Tucker, T. J., and Kling, A. Differential effects of early and late lesions of frontal granular cortex in the monkey. *Brain Res.,* 5:377-389, 1967.

THE EFFECT OF AGE ON THE OUTCOME OF CENTRAL
NERVOUS SYSTEM DISEASE IN CHILDREN

ERIC H. LENNEBERG

Professor of Psychology and Neurobiology,
Cornell University

As students of behavior we have always been aware that organisms differ from most machines in that they have an experiential history. Perceiving and responding have a cumulative effect upon further propensities for behaving, and so probabilities for actions are never quite the same throughout an animal's lifetime. There is yet another difference between living organisms and machines. In the case of artifacts that we build—mechanical devices or electronic equipment—it is easy to make a distinction between the physical gadget, on the one hand, and its behavior, on the other. This is most obvious when we build a machine that we allow to stand idle; in this case it does not behave at all. Living organisms, however, always behave. There must always be cells secreting or absorbing substances, muscles must maintain tonus, organs must contract or pulsate, and neurons must send out impulses. A standstill marks the end of existence of a living organism. It is not a tautology to say that organisms must be active in order to remain live organisms. It is theoretically conceivable that the notions *experience* and *behavior* could be defined in such a way that they describe states and functions that are totally separable from the biological constitution of the organism—that they have nothing to do with the construction and maintenance of the machine. In fact, this was precisely the hope of the fathers of behaviorism. Today, however, with our increased knowledge of the relationship between brain and behavior we take a more skeptical view of this possibility. In the present article I wish to emphasize the intimate relationship between aspects of behavior and aspects of physical growth and development. The two constantly influence each other and the intimacy of the relationship is dramatically seen when we make certain comparisons of the effect of environmental or physical abnormality in childhood with the effect of identical conditions in adult life.

From this type of consideration we obtain a more dynamic view of the capacities of organisms. There is no such thing as a stable organism with unchanging properties. The study of behavior must take into consideration the changing propensities of the individual, particularly during the period from conception to

147

maturity; and even after puberty the stability is only relative; and further changes occur until the most dramatic one of all is reached: death.

Further, species differ in their natural life-histories. The degree of maturity at birth varies a great deal among mammals. Puberty is reached at different relative points in the course of a given animal's development, and average longevity and the point of decline of capacities with age are also species-specific. Because of this variation one must be very careful in the degree of cross-species generalization. Even individual organs and functions have different relative maturation curves.

Man is rather unique among primates with respect to his maturational history, as illustrated in Figure 1. These maturational peculiarities make it essential that we do not generalize from the sort of experiments that have been reported on earlier in this symposium and blindly apply the knowledge gained there to man. It is imperative that we supplement this knowledge by observations made directly on our own species. But man is not an ideal experimental subject because we cannot do the same kind of experiments as we can do with animals. However, the situation is not as hopeless as is sometimes claimed. The history of physiology shows that there are certain types of experiment that are feasible, though I admit they are much more limited in scope than the ones we allow ourselves to do on animals. On the other hand, knowledge about humans is made available through

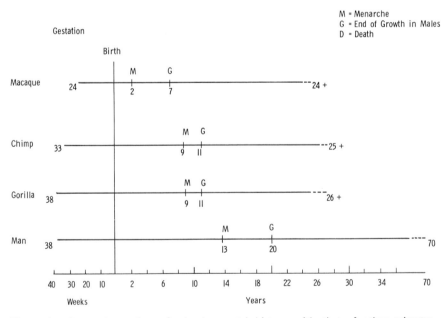

Figure 1. Comparison of man's developmental history with that of other primates. (Based on Schaller, 1963, and Schultz, 1956.)

a much more detailed study of disease and the possibility of more accurate description of the consequences and reaction to disease. Patients can tell us what they feel. Furthermore, there is no animal about which we have as accurate information about what it should be doing—how, when, and where—as we have for man. Thus the availability of well controlled lesions and functional abnormalities is limited in man, but the cases that happen to come to our attention and that have "interesting" disturbances in their wake can be studied with a degree of precision and detail that is simply unattainable in the case of animals.

With this preamble let us look at a small selection of abnormal conditions in man and see what conclusions we may draw from them with regard to the development of the central nervous system.

THE EFFECT OF BEHAVIOR ON SOMATIC DEVELOPMENT

Abnormal environments often elicit abnormal behavior in children, and this, in turn, may have further deleterious effects upon the emotional atmosphere created by the parents, resulting in a vicious circle of cause and effect in which the growing child is the paramount victim. Recently an important study was published (11, 12) that illustrated the chain as shown in Figure 2.

Thirteen children were studied over a number of years. All of them had been admitted to a hospital because of dramatic retardation in growth. Table 1 shows the growth failure in terms of height standards. Each one is markedly below the first percentile of the height distribution for his chronological age. All of these children had bizarre personal histories that set them apart from the more commonly encountered pituitary dwarfs. The thirteen cases came from eleven homes (there were two pairs of siblings). Interviews with the parents and social service home visits revealed an abnormal family background in every case. Five of the fathers had left the home, and the remaining fathers were described as

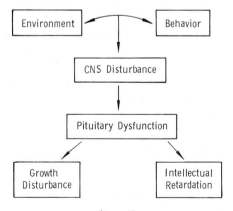

Figure 2.

TABLE 1. GROWTH MEASUREMENTS OF CHILDREN WITH HYPOPITUITARISM
DUE TO EMOTIONAL STRESS—INITIAL VISIT[a]

Case Number	Chronologic Age (years)	Height Age (years)	Bone Age (years)
1[b]	3.3	1.8	1
2	3.8	2.0	2-3
3[b]	3.9	1.2	3
4 ♀	3.9	2.6	3
5	4.4	2.0	2
6[b] ♀	5.1	2.8	5
7[b]	5.1	2.0	2.5
8	5.3	3.5	4
9	5.8	3.0	4
10	6.9	3.5	3-4
11 ♀	7.2	4.0	6
12	7.8	4.5	4
13	11.5	5.3	6

[a]Based on Powell et al., (1967).
[b]Siblings (cases 1 and 6, 3 and 7) belong together.

having a bad temper or maltreating the children; in two cases there was marked marital strife; one mother was an alcoholic and another a psychotic. There was no family without a history of either desertion, violence, alcoholism, marital anomalies, or several of these. While these conditions are hardly *that* unusual in our sociological fabric (and therefore cannot be considered the principal cause for the children's physical difficulties), the patients' own behavior patterns were. All of the children had polydipsia and polyphagia that manifested itself in characteristic but unusual ways. Six of the children were frequently found to drink from the toilet bowl; some drank dishwater; others drank from puddles of rain water or they drank stagnant water out of old beer cans. Many arose during the night to drink water; one of the children had been locked up at night to keep her from getting up repeatedly. Most of the patients ate two to three times as much as their normally growing siblings. There were stories of children eating a whole jar of mustard or mayonnaise, a package of lunchmeat, a whole loaf of bread, corn flower from a box, and seven eggs at one sitting. Some children ate the cat's food and every one of the children scavenged the garbage cans. Many of the children would frequently gorge themselves with food until their stomach would swell and all of them vomited frequently. These aberrations were reminiscent of hypothalamic disorders. There were also insomnia, temper tantrums, stealing, and standing around with vacant stares.

Eight of the thirteen children had psychometric testing, and all were considerably behind their norm. Speech was retarded in all but two. In these exceptions, interestingly enough, the physical symptoms developed only *after* speech had been established during the third year of life. Gait was also retarded in eight out of thirteen.

In addition to these behavioral aberrations the following physical symptoms were noted: the birthweight and neonatal history were normal in all patients; commonly the onset of the condition was fairly sudden and was accompanied by markedly foul-smelling and bulky stools; symptoms developed in the course of the second year, except in two cases where the disease made its appearance suddenly at 3½ and 4 years of age.

All children were given extensive physical and clinical tests at the time of their admission, and there was evidence in all of them of hypopituitarism (Table 2); none of them gave a clinical picture of malnutrition and this, plus other details of the clinical studies, made it possible to rule out a malabsorption syndrome.

The treatment of all of the children consisted of temporary removal from the home environment and admission to a convalescent home where they were given no medication of any kind and no psychotherapy. In every case the behavioral and physical symptoms ceased within a few days of environmental change, and all the children began to grow at considerably faster rates than is normally seen. This catch-up growth is illustrated in Figure 3.

During and toward the end of the hospitalization period, which lasted from a few months to a year (several cases were followed for longer periods), endocrine studies were continued; there was evidence for absence or paucity of growth hormone before admission and for a restoration of hormonal balance afterward.

TABLE 2. EVIDENCE FOR HYPOPITUITARISM BEFORE GROWTH[a]

Evidence	Number of Children
Short stature	13 of 13
Retarded bone age	10 of 13
Low PBI	1 of 13
Low ^{131}I uptake	3 of 13
Low base-line corticoids	9 of 13
Abnormal result on metyrapone test	12 of 13
Insulin sensitivity	2 of 11
Decreased growth hormone	6 of 8

[a]From Powell et al. (1967b).

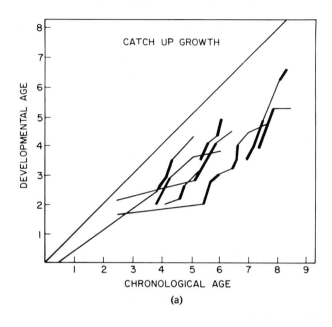

(a)

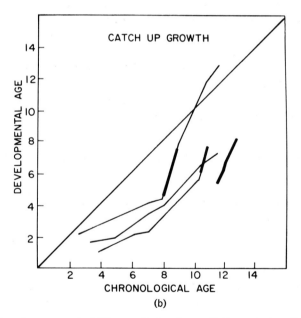

(b)

Figure 3. Growth curves of children suffering from pituitary dysfunction caused by environmental abnormalities. Change of environment *without* other treatment induced the catch-up growth shown by the heavy lines. *(a)* Younger patients. *(b)* Older patients. (Based on Powell et al., 1967; reproduced by permission.)

152

Catch-up growth could thus be accounted for by the sudden release of growth specific pituitary hormone. The authors of this study conclude their report thus:

"Much evidence has been accumulated that secretion of pituitary tropic hormones is influenced by the hypothalamus. Specifically in regard to somatotropin secretion, lesions of the anterior hypothalamus in monkeys and rats prevent the usual growth-hormone depletion of the anterior pituitary body with insulin induced hypoglycemia. In addition, hypothalamic extracts have been found to stimulate the release of growth hormone. We can only postulate that emotional disturbance in these children may have had an adverse effect upon release of pituitary tropic hormone via the central nervous system."

These cases illustrate how perceptual and behavioral factors may have effects on central nervous system functions. When this happens during childhood, critical periods may be involved during which growth and development take place. If the condition is treated in time, the adverse effects are largely reversible; but if treatment is delayed beyond the time of puberty, stunting or retardation is irremediable. Clearly, the particular reaction to stress illustrated by these cases must be regarded as an oddity. The implication of this presentation is not that psychological stress at this age-range necessarily produces pituitary dysfunction. Stress is such a nondescript stimulus that it is impossible to make unfailing predictions about an individual's physical response. There are different types of stress and different types of response to each. Most of the children studied here had siblings, but in only two instances did more than one of the children in the same family react by physical retardation. If the other children in the households had been studied we might have found incidences of migraine, or asthma, or colitis, or obesity. Phenomena of this sort are well known in psychosomatic medicine. The symptoms described here were chosen for presentation in this context for two reasons: first because they might have had irreversible effects upon the development of the children, and second because of the exceptionally careful documentation of the condition.

THE EFFECT OF SOMATIC ABERRATIONS ON BEHAVIOR

Of the many examples of this type of causal relationship I present only one: hypothyroidism. The cause and effect sequence may be represented as in Figure 4.

The conditions of cretinism are well enough known to make an elaborate description of symptoms unnecessary here. Instead we will concentrate on two aspects of the disease: (a) the behavioral consequences of the disease, and (b) their relation to the age of onset of the disturbance and treatment of symptoms.

Dysfunction of the thyroid gland produces, among other physical symptoms, the following psychological abnormalities: dulling of the intellect, drowsiness,

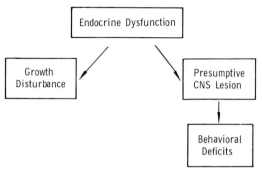

Figure 4.

apathy, anorexia, amnestic difficulties, inability to make inferences, motor clumsiness; the voice is hoarse, low pitched, and monotonous; speech is slow, often with a nasal quality, and as inarticulate as during a state of drunkenness; stance and gait have a characteristic stiffness in which limbs are slightly flexed and the trunk is bent forward.

At birth two variants of the disease are seen. One is restricted to iodine-poor, underdeveloped areas, where endemic cretinism is common. The other is congenital athyroidism, which occurs spontaneously. Althouth this distinction is not of great consequence for the clinician, it is of particular interest to us here. The first type is related to the mother's poor thyroid function, which affects the embryo during earliest fetal life. The infant at birth has already suffered from the hormonal insufficiency; he is thus born with a deficit that is most marked with respect to his intellectual potentialities at birth, whereas his physical symptoms (myxedema and dwarfing) are much less pronounced. The other type has had a normal fetal life, because the mother's blood supplied him with sufficient hormones throughout gestation. After delivery the hormonal absence has an immediate effect, which manifests itself physically and behaviorally. But in this case the cerebral damage occurs at a later formative age than in the other type. This difference has its consequences upon prognosis, to be discussed presently. In addition to the conditions described, there is also an acquired form of hypothyroidism; this may occur at any age.

The age of onset of the disease is of paramount importance for the outcome of the condition. In Figure 5 we see the clinical picture of two patients with hypothyroidism. In one instance the disease was congenital and untreated; in the other the patient became athyroideic at about 45 years of age. When the disease strikes at an early age, not merely is growth retarded, but behavioral and psychological development are also irreversibly affected. The black-box psychologist who wishes to explain behavior entirely by environmental treatment may be inclined to attribute the inferior intellectual status of the congenital and

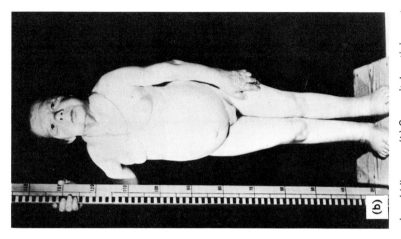

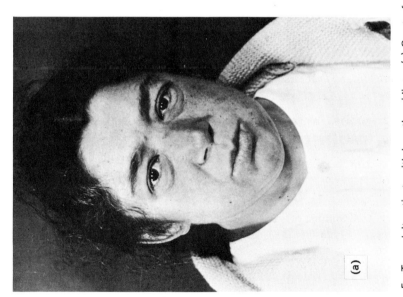

Figure 5. Two adult patients with hypothyroidism. (a) Onset of the disease was in middle age. (b) Congenital cretinism untreated during childhood. (a) Reproduced from Labhart, 1957 by permission; (b) through the courtesy of Professor Labhart.)

untreated cretin (in contrast to the patient who acquired the disease in adulthood) to his abnormally restricted experience. The adult patient stopped learning much later than the congenital one.

There are compelling reasons, however, to dismiss this hypothesis. There are indications that the central nervous system is affected in very different ways during childhood and during adult life. The two conditions result in very different physiques; hypofunction in childhood leads to irreversible developmental arrests, the severity of the deficits varying in direct proportion to the duration of lack of hormone treatment during childhood; reinforcement therapy or other forms of environmental manipulation do not change the basic capacities of the cretin.

The importance of age for prognostication cannot be over-estimated. In children the condition is totally curable only in the case of congenital athyroidism and only when symptoms are identified within a few weeks after birth and therapy is initiated immediately (see Table 3). The infant with endemic cretinism can also be treated with advantage but his intellect can never be restored to normal limits, presumably due to the prolonged prenatal brain damage sustained. The prognosis for his physical development is good, however. Interruptions in thyroid therapy during the period of potential growth tend to be detrimental. When therapy is reintroduced, limited catch-up growth may occur, but intellectual functions are not likely to be fully restored, the extent of the improvement depending on the severity of the condition and the length of therapeutic neglect.

Figure 6 shows the developmental curves of a child whose hypothyroidism had been allowed to pass untreated until age four. Observe the rapidly increasing rate of growth. Nevertheless, it was evidently too late for attaining normal stature.

TABLE 3.   MENTAL ATTAINMENT IN SEVERE CONGENITAL HYPOTHYROIDISM[a]

Age at Which Treatment Was Begun	Number of Cases	90 or More (percent of cases)	70-89 (percent)	50-69 (percent)	50 or Less (percent)
		Highest IQ Attained			
0-6 months	22	45	27	10	18
7-12 months	7	29	42	0	29
12 months and over	22	0	41	18	41

[a]From Smith et al. (1957).

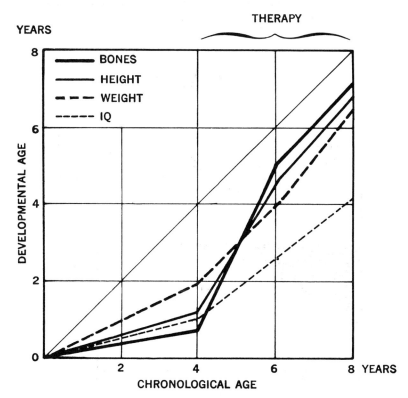

Figure 6.   Developmental acceleration in response to hormone therapy of a patient with congenital hypothyroidism. (Based on Labhart, 1967, and reproduced by permission.)

Intellect also improved under therapy, but its acceleration was markedly less than that of physical growth. The graph shows that IQ tends to fall behind. In fact, the intellectual deficit becomes even more dramatically noticeable at later stages of cases like these. Despite the continued treatment, the children are condemned to intellectual moron levels for life.

A severely feeble-minded cretin who has not been treated by age ten cannot be ameliorated in any way by later treatment. Administration of thyroxin or related drugs may, in fact, merely produce a degree of agitation that is undesirable for reasons of management; such patients are therefore left untreated by many physicians.

The effects of hypothyroidism on mental development are most severe during earliest infancy. If the disease develops some years after birth and the child has previously had the benefit of well-regulated thyroid function during his most critical developmental period, his outlook for a reinstatement of good mental capacities is much better. This is dramatically demonstrated in Figure 7.

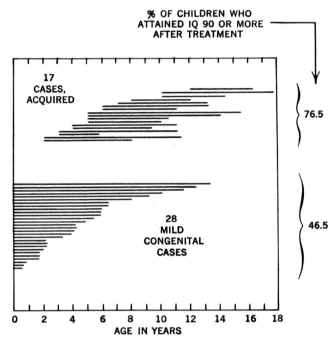

Figure 7. Different periods of treatment of children with hypothyroidism. Above: patients with acquired thyroid disease. Below: patients with congenital hypothyroidism. Less than half of the congenital cretins attained normal IQ's, whereas three-quarters of the children with acquired hypofunction had an IQ of 90 or better after treatment. (From Smith et al., 1957.)

The prognosis is entirely different in the case of hypothyroidism acquired after maturity. Although mental deficits occur regularly, they are easily improved with therapy and the outlook does not appear to be correlated with the length of lack of treatment. Since hypothyroidism is compatible with life, patients were described at the beginning of the century who had had the disease for as long as 20 years. In one such case treatment was begun in the fifth decade of the patient's life. Even though he had had the disease for many years, in this case he was restored to normality and he lived to be a hoary 90 (5).

FURTHER EXAMPLES OF CRITICAL PERIODS

Further illustrations will now be reported briefly. I have arranged the diseases in order of the developmental phase during which they affect the embryo or growing infant. It is most likely that in each case entirely different mechanisms are involved, even though the outcome in each instance tends to be feeble-mindedness. Actually, feeble-mindedness is a secondary consequence; the primary defect that blocks mental development must be assumed to be very different in these various conditions, because each one has its own characteristic clinical picture. In the present context it would not be warranted to present all of the physical signs and symptoms of each disease.

Mental Defects Resulting from German Measles during the First Trimester of Pregnancy. Contraction of the disease from the second to the third month of pregnancy appears to have the worst consequences on the morphology of the embryo. During the fourth month the chances for deformations rapidly declines, and later during gestation there appear to be no ill effects. Among the most common defects are ocular deformities affecting the retina, the formation of the lens, or the growth of the entire eyeball. Hearing defects also occur. In most instances there is at least mild mental retardation, though exceptions are not uncommon. Occasionally, severe feeble-mindedness is the most prominent sequella. The head circumference is usually in the low percentiles. Frank micro-cephaly has been quite frequently described in the literature. One autopsy report of a 13-month-old child has been published (3) in which the brain was found to have a defective development of the cortex, the internal capsule, and the cere-bellum, and there was a complete absence of the corpus callosum.

Retrolental Fibroplasia. This disease is rapidly disappearing, because it is now known that it was induced by the physician. If premature babies with a birth-weight of less than 3 to 4 pounds are exposed to high oxygen concentrations—such as are produced in oxygen tents used in the clinic for conditions causing respiratory insufficiency—a deformation of the retina and vitreous may result, which may render the baby blind for life. In addition, various cerebral defects may be associated with this, resulting in congenital cerebral diplegia and convul-sive disorders. Other developmental defects of the brain are also suspected (2) but have not yet been proven. This is an example of late embryonic sensitivity. It is interesting that oxygen treatment, which at later ages is perfectly benign, can have such bad effects if it occurs at just one particular time during development.

Kernicterus and Hyperbilirubinemia. Irreversible damage to the brain may be caused at birth through extreme degrees of jaundice in the neonate period. This condition is mentioned in the present context because severe jaundice at later stages of development and in the adult have no such irreversible effects on the brain. The most common cause of pathological levels of jaundice in the neonate is erythroblastosis fetalis (RH incompatibility). The coloration is due to the circulation of the pigment *bilirubin*. In blood incompatibilities it results from an accumulation due to a high degree of hemolysis, which releases the pigment. Other types of jaundice may have similar effects; the bilirubin concentration in the blood, for example, may be due to an enzymatic defect (bilirubin transferase) that blocks the normal elimination of the substance.

The word *Kernicterus* indicates "yellow-green stain of the basal ganglia" in the brain. These structures (lenticular and caudate nucleus) are most severely affected, but medullary nuclei, the optic nerves, fiber tracts of the spinal cord, the hippocampus, and the dentate nucleus of the cerebellum may also show the stain. Whether the cerebral damage is due to the presence of the stain is not certain. However, deep jaundice in the adult does not stain the cerebral structures and does not result in any permanent central nervous system symptoms. On the other hand, it is possible that the brain damage in the neonate is the result of secondary factors such as myelinization defects, secondary vascular lesions, or hypoxemia. At any rate, deep jaundice at this age and its associated brain damage leads to cerebral-palsy-like states. It may be accompanied by muscular rigidity, mental deficiencies, convulsions, and athetosis. A fairly common and rather specific disorder has to do with the capacity for the perception of speech patterns. This is often referred to as cortical deafness. The term is misleading because it is not certain that the hearing threshold for tones is elevated in these patients nor is there any evidence that there are cortical lesions. Children with kernicterus can often be observed to respond to artificial sounds. But they show a peculiar inattention to language stimulation and frequently fail to develop language at the expected time.

Phenylketonuria. The maximal brain damage brought about by this disease seems to occur during the third year of life. It belongs to the family of diseases known as inborn errors of metabolism. It is a genetic anomaly, inherited as a recessive trait. Heterozygous parents can be detected by analysis of the urine. Specifically, the patient suffers from hepatic enzyme defect of phenylalanine oxydase, which normally transforms phenylalanine to tyrosine. The absence of the enzyme leads to an accumulation of phenylalanine in the blood and in the tissues in general. A cumulatively abnormal chemical environment created in the brain causes mental deficiencies. At birth there are no abnormalities, and children with the disease (demonstrable by showing phenylperuvic acid in the urine) are free of symptoms for days and sometimes even a few weeks after delivery. But as phenylalanine begins to accumulate, the pathology becomes evident and brain damage begins. Treatment consists of a special phenylalanine-restricted diet. Clinical experience with this diet and its usefulness at various ages suggests that the brain damage reaches its peak between the second and third year. After that the adverse effects level off rapidly and seem to have no further detrimental consequences upon the brain. Table 4 shows the effects of the diet upon IQ. The deterioration of the IQ figures is not to be attributed merely to a progression of the disease but is, in part, an artifact of the mode of reporting on intelligence. Early brain damage sustained in this disease and its consequences upon behavior are irreversible (1).

TABLE 4. DIFFERENCE OF AVERAGE IQ BETWEEN TWO
PSYCHOMETRIC TESTS; ALL SUBJECTS HAVE
PHENYLKETONURIA; SOME WERE TREATED BY DIET[a]

	Differences Between Average IQ Scores; Tests Were One Year Apart	
Age Group	Untreated $N = 19$	Treated $N = 47$
Under 5 years	−8.3	+10.5
5-14 years	−4.7	+1.0

[a]Based on Bickel and Grueter (1960).

THE NATURAL HISTORY OF DISORDERS DUE TO CEREBRAL INJURIES

So far we have discussed only systemic diseases that result in hard-to-define lesions; most of them are on a cellular or microscopic level. Let us now consider the outcome of more specific structural lesions with regard to age. Caution is in place here. The field of developmental neurology is so vast and complicated, and the unexplained problems are so awe-inspiring, that any attempt to discuss it comprehensibly in a few pages here must strike the expert as sheer megalomania. In fact, it is not my intention even to do so much as to give a survey of the field. Instead, I shall choose a few phenomena that are well illustrated by a study of children with central nervous system disease, in order to show some of the dimensions of the "great unknown": the development of the brain and of its function.

When lesions are incurred in the mature brain, there is a certain sequence of events, such as transgression of fluids into the interstitial spaces, softening of tissue, cavitation, necrosis, and gliosis, but in most instances the lesions remain relatively confined to specific structures. However, lesions incurred during the prenatal or early postnatal period are often followed by much more widespread pathology, because tissues are still interacting in morphogenetic and developmental growth processes. Norman (8) described a number of autopsied brains of children who had had hemiplegia due to birth injuries. The lesions were presumably caused by vascular compression at birth. In the adult patient, vascular accidents causing hemiplegia are very common, and many pathological investigations have been carried out on their brains. To compare the pathologies of the children with those of the adults' brains is complicated by the fact that the vascular insufficiencies are not identical. This precludes a narrow comparison of the exact topographic extents of the lesions. However, a juxtaposition of the pathological processes involved shows that in the children a series of far-reaching alterations and consequences may follow the initial injury, months or perhaps even years later; there are no good parallels for this in the adult cases. An outstanding example is the disturbance in myelinization seen in the children's brains. Both hypermyelinization and plaques of poor myelinization are commonly seen in birth-injured brains. Other examples are sclerotic microgyria, linear scars of dense, fibrous glial cells, groups of fat granule cells, and calcification of debris of destroyed tissues. Brain injuries in prematurely born infants may have even more widespread consequences, particularly if structures are involved that give rise to migrating cells, such as the cerebral ependyma. Neurogenesis apparently is just about complete in man at term, although glial cells continue to be formed and to migrate for a considerable time after birth. Local interference with this process due to pathological anatomy may be responsible for dysplasia, which, in turn, may have secondary consequences in more remote areas of the cerebrum as a whole. Presumably, asymmetry in volume and geometry between the hemispheres and ventricles are examples.

Concomitant with the difference in the nature of the lesions between children's brains and adults' brains, there are also important differences in the ensuing symptomatology. Let us discuss symptoms under the following four headings: (*a*) motor; (*b*) sensory; (*c*) language; (*d*) intelligence.

The motor disturbances due to cerebral birth injuries cannot be fully understood without an appreciation of the physical processes of growth in general. Most relevant here is the as yet unexplained relationship between cerebral injuries and skeletal growth. Lesions that involve the cerebral cortex, particularly lesions of the parietal lobes, have a retarding effect on the growth of the long bones of the contralateral limbs. Arms and legs are equally likely to be affected, although the stunting is usually more obvious in the arms. Figure 8 is an illustration of this. Children who have incurred lesions of different size and distribution in both hemispheres may suffer stunting on both sides of the body but in different parts; for example, a short right arm and a short left foot. The hypoplasia may affect muscles independently from bones, and vice versa, or both tissues may be affected simultaneously and proportionately. A number of investigations have been made to see whether the underdevelopment is due to disuse, but most authorities now agree that this is not the case and that the growth disturbance must be regarded as a trophic disorder (4). This observation stresses once more the intimate relationship between various tissues and their function during morphogenesis.

The study of hemiplegia furnishes good comparisons between the consequences of CNS lesions during infancy and during adult life. We shall consider only cases in which the injury occurred not later than the neonatal period and consists of a fixed lesion. The most dramatic difference between infancy and adulthood in this respect is that in infancy the lesion produces no measurable effect whatsoever for at least the first three months—often much longer. All four extremities move

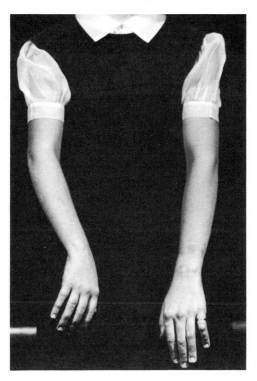

Figure 8. Stunted growth of right arm due to left-sided cerebral lesion at birth. (From Paine, 1960, reproduced by permission.)

well and symmetrically at first, and growth is as yet entirely unaffected. In fact, children without any telencephalon, that is, hydranencephalic infants (demonstrable by transillumination), may show no behavioral signs of anomaly whatsoever during the first 3 weeks. Presumably, the behavior that we see at this time is under the control of diencephalic and mesencephalic structures only, the cortex being still too immature to have any effect. When higher centers, probably the cortex, begin to exert an influence on the behavior of the normal infant—at about 3 months—the first signs of abnormality make their appearance in the hemiplegic child. But they are still so inconspicuous that they frequently escape the parents' attention and it takes a skilled observer to demonstrate the emergence of signs. Tizard (18) lists three aspects of symptoms at this age: (*a*) deficiency of movement or posture; (*b*) alterations of muscle tone; and (*c*) preservation of primitive reflex patterns.

The motor deficiencies are first noticeable in the arms and hands; the earliest time for this is about the fourth month, but it occurs more commonly in the sixth month. Legs can be demonstrated to be moving slightly differently from each other at not earlier than 9 months, more commonly at 10 to 12 months. The affected arm is held more quietly and is never used for reaching. Objects put into that hand are ignored. The baby, in fact, does not seem to be paying attention to what is done with the affected limb. The mother may notice, while she is dressing the child, that the arm is less pliable, that it puts up a mild resistance. This is the beginning of spasticity or muscle tonus abnormality. While the fingers on the good hand begin to move independently, the bad hand is usually clenched into a tight fist. The grasp-reflex disappears on the unaffected side by the third month but continues to be elicitable on the affected side for a much longer time. Similar observations may be made on the tonic neck reflex.

It is of particular interest that hemiplegia in early childhood is not a paralysis in the sense that it is in the adult. While the affected limb is never used and apparently lacks the synergisms for reaching or grasping, it is occasionally seen to move either spontaneously or in response to a noxious stimulus. When it does move, however, it seems to move in a somewhat unpredictable manner. During neurological examination the patient may ignore a pinprick or pinching at one time, but at other times a response may be quite unequivocal.

The motor defect in the lower limb is usually noticed only at the time the baby begins to walk. From Table 5 we see that this milestone is frequently attained at the normal age. Abnormal forms of gait will invariably develop, but they differ markedly from the abnormal gait seen in the adult hemiplegic. Naturally this is due partly to the frequent growth disturbances that accompany the natural history of congenital hemiplegia. But it is also related to more general growth and development processes. The stunting as well as the spasticity force the child to assume abnormal postures to maintain his balance and to counteract gravitational pull in directions that diverge from the normal body axis. In many children contractures develop and bring forces to bear upon bone formation that contribute to further skeletal distortions. All of this affects the whole body not only in morphology but also in the formation of motor coordination patterns.

TABLE 5.   COMMENCEMENT OF WALKING IN
HEMIPLEGIC CHILDREN[a]

Onset of Gait at Age (months)	Number of Cases
11-13	19
14-18	57
19-24	28
25-36	7

[a]Based on G. Lyon (1961).

While spasticity is the first symptom and paralysis a much later one, the most commonly seen symptom of the more mature cerebral-palsied child—athetosis—is the last one to emerge. As a rule it is not seen in children before they are 3 to 4 years old.

In summary, one may say that the child with a perinatal cerebral injury only gradually "grows into his symptoms," and that both lesions and symptoms have their own ramified consequences, often affecting distant structures years after the primary injury.

Turning now to the question of sensory deficits, we may be much briefer than in our exposition of motor problems. Sensory testing in children is always difficult but is more so in defective children, especially when they are irritable or mentally retarded, as some of the children with hemiplegia are. Unfortunately, the sophisticated equipment and the training periods necessary for accurate investigation of sensory deficits are not yet available in children's clinics. Paine (10) has made some important observations. He agrees with other authorities that injuries affecting area 17 of the cerebral cortex of either hemisphere produce the same field defects as are seen in adults. There are, however, many problems with respect to pattern perception and the exact nature of the field defect which we cannot review here. Hemianopias and scotomata are not very frequently seen as a consequence of birth injuries, though they do exist. More common are hearing defects. For reasons that are not clear, the perception of high-frequency sounds is said to suffer more than that of low-frequency sounds. It is my own clinical impression that the perceptual problem is not clearly confined to one range in the sound spectrum. Instead it seems to be sound *pattern* perception that suffers most. Some workers equate this condition with receptive aphasia of the adult patient. There are difficulties with this interpretation. The physical signs that accompany adult receptive aphasia have a different appearance and distribution and usually point to different cerebral pathology. (More about this in the following sections.)

Somatosensory losses have also been described including touch, pain/temperature, position/vibration sense, two-point discrimination, and stereognosis. These symptoms, compared to motor symptoms, speech, and IQ, diverge least from the adult symptomatology. Perhaps this statement will have to be revised when better testing methods become more readily available in the clinics.

In this sampling of behavioral consequences, language takes a special position. In adult right-sided hemiplegias language disorders are extremely common. In congenital hemiplegia, however, they are relatively rare. The symptoms seen in the adult are virtually never seen in children, except for dysarthria, and even in this case articulatory disorders of hemiplegic children have very different acoustic properties and seem to be caused by motor abnormalities that differ from those of the adult at least as dramatically as the gait in children differs from the gait in adult patients. The side of the congenital hemiplegia and the side of the lesion in the brain make no difference in this picture. In order to understand the full implications of this finding we must digress for a moment and also consider traumatic lesions acquired later during childhood.

Unilateral cerebral lesions incurred before age two do not implicate subsequent language development. After age three left hemisphere lesions in the frontoparietal area cause the patient to lose language temporarily, but soon it is fully reinstated. This impunity lasts until approximately age 10 to 14 years. At this time and thereafter aphasic symptoms rapidly become more frequent and in about 30% of all cases are irreversible. From this we must conclude that the cerebral correlates for language only become organized well after birth, probably around age two. For another eight to ten years after this time there is still little specialization with respect to topographical differentiation. It is only after the brain has reached full physical maturity, roughly around the time of puberty, that the major language functions are locked into place, so to speak. If this theory is correct, and the weight of available evidence supports it [see Lenneberg (6)], then language is unique with respect to the extraordinarily late time of its embryological determination. While localized damage to the central nervous system before or around birth can block the proper development of perception and motor skills, and thus irremediably distort these aspects of behavior, localized damage to presumptive speech areas does not have any effect until relatively late in childhood.

The natural history of aphasia-producing lesions is in strange contrast to the natural history of intellect-limiting cerebral insults. Both types of behavior—language and reasoning power—are biological developments that are unique to our species. One therefore may expect some parallelism, whereas, in fact, a biological study of the two reveals them to be quite separable phenomena. Behaviorally this is confirmed by the lack of correlation between the two. Congenitally deaf children who have no language before entering school perform as well on totally nonverbal intelligence tests as their hearing contemporaries. On the other hand, children with an IQ too low to allow them to go to school (say IQ 50 at age 12) are frequently quite proficient in language.

The dissimilarity between language and intelligence is also reflected in the aftermath of cerebral trauma. The destruction of as much as 30% of cortical tissue in either hemisphere appears to be compatible with fairly normal intellective functions. This seems to be true of traumatic lesions acquired in infancy (19) as well as traumatic cortical lesions sustained later in life (17). In the adult, this observation is independent of the presence or absence of aphasia (20). The limitation of intelligence discussed earlier in connection with a number of systemic diseases is therefore not due to the destruction of islands of cortical tissue but to interference with more general cellular processes, hampering interaction between many parts of the brain. The skewed distribution curve of the incidence of various IQ's in birth-damaged children, as shown in Figure 9, is probably due

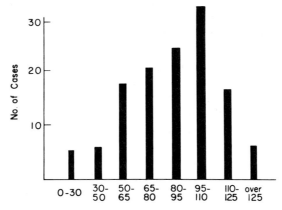

Figure 9.   Frequency distribution of IQ in a group of 140 children with infantile hemiplegia. (From Woods, 1961.)

to a sampling artifact. Frequently this group includes children who have suffered anoxia at birth with subsequent widespread—selective and scattered—cell destruction in addition to the trauma that rendered them hemiplegic. It is this subset of the sampled population that is probably responsible for the left-sided skewing of the distribution curve.

CONCLUSION

This sketchy survey of some neurological diseases in childhood makes us inescapably aware of the unity between behavioral capacities and biological, especially neurological, mechanisms. If we wish to understand behavior we can no longer afford to regard it merely as an input-output function. It is especially facts such as those reported here that force us to think of behavior as a phenomenon inseparable from the machine that behaves. Many of the environmental factors that have been known to be stimuli for behavior are now seen to be also stimuli or factors that affect the very construction of the machine that behaves. The best example of this is the adverse social climate which, in children with a certain organic disposition, may affect pituitary function through influences from the hypothalamus, which in turn is responding to emotional stress mediated through perceptual and cognitive cerebral mechanisms. The result is not only a stunting in growth but also and concomitantly a stunting in intellectual development. We know from a wide range of experience that this condition becomes irreversible if protracted into adulthood. Hypothyroidism, on the other hand, begins with an error in endocrine function that has an immediate effect upon the evolution of intellectual capacities.

In these cases, as well as in the case of phenylketonuria and Kernicterus due to blood incompatibilities, it is obvious that the diseases come about through genetic dispositions. A decade ago learning-theorists seemed to believe that a demonstration of genetic background was a signal for them to exit from the stage of science and to let the biologist enter—that their contribution to the drama of behavior had ended. Today many psychologists know better. Genes set the limits to everything in life, including behavioral propensities, but genes do no more than say: "Given condition A, then cellular response B." The condition is always environment—either the internal environment (both the cytoplasm surrounding the cellular nucleus and the ambient surrounding the cell) or the external one, and frequently both. Thus genetics does not turn psychologists out of a job.

Furthermore, the demonstration of or search for genetic controls does not mean, as is still constantly assumed by behavioral scientists, that "something is preset from the start." A survey of conditions that act at very specific times upon the developing organism illustrates that development as a whole must be thought of as a long chain of sensitive states or periods during which it is critical that certain aspects of the environment are just right; that the organism is provided with that particular nourishment or those particular stimuli that it needs during that particular maturational phase for further growth and development. On the other hand, all developmental stages are subject to stimuli that are harmful and can interfere with proper development of form, function, and behavior during one stage but are harmless at all other times. We now know that genes are "turned on and off" throughout development; and this is accomplished by circumstances partly created by the growing organism itself and partly by exogenous factors to which the organism opens itself up or prepares itself for at one and only one developmental stage. This is the reason why misfunctions, trauma, or infections may have far-reaching consequences at one period, but relatively few at another. The same seems to be true of various environmental deficiencies. Undernourishment is tolerated rather well in the adult but may have irreversible effects upon intellectual capacities in the infant (9, 16).

There is a vast number of hereditary CNS diseases that suddenly cause various cells and tissues to degenerate. All of these diseases strike at specific stages of human development. Some manifest themselves at the toddling age, some in the early teens, some in young adulthood, others in middle age or senescence. Suddenly cells fail to function; the patient may experience motor discoordinations, have seizures, mental dysfunctions, change of personality or emotional instability. His behavior may revert to that of a small child, or he may have delusions. The pattern of inheritance may clearly point to genetic causes, but this merely says that in the long chain of propensities that genes create, a link is missing and the continuity of smooth function is interrupted.

Ethologists have introduced the notion of the critical period to the study of behavior. There are psychologists who find their theoretical edifice threatened by the demonstration that animals go through periods at which they have heightened sensitivities, so that patterns of behavior are established very rapidly and are highly resistant to extinction. But if we look at behavior from a biological point of view, we should be surprised if we did *not* find critical periods. The developmental history of organisms, both of the machine and of its behavior, is one long chain of phases in which one or another set of factors is of critical importance. Deviations from the optimal conditions may have dire consequences. The survey of neurological diseases in childhood teaches us just that. It reminds us of the most basic embryological principles and shows how they are applicable to an understanding of the development of behavioral capacities.

For instance, a lesion in the internal capsule has very different consequences in the neonate than in the aged. Cells, tissues, and organs are differentiated at a certain time and shortly thereafter reach finality of form and function; a lesion incurred before this final stage may, in some cases, be compensated for by internal readjustments or compensatory growth. But once the terminal stage of differentiation has been reached, compensation is no longer possible. Not all structures reach this final stage of differentiation at the same time. An abnormality established in one part of the brain may affect other parts to a degree that depends on their stage of differentiation. In embryology one calls the capacity for adjustment to structural abnormalities *regulation*, and the final stage of differentiation is called a process of *determination*. A fine example of regulation and subsequent determination is given by left hemisphere lesions in the frontoparietal lobes. Determination with respect to language takes place only at the beginning of the second decade. Earlier lesions are compensated for; later lesions frequently leave permanent deficits.

These sketchy remarks will, I hope, foster a renewed interest in the study of behavior as a biological phenomenon. Throughout childhood, behavioral capacities are constantly changing in accordance with a program that is genetically encoded and that determines the stimuli to which the growing organism shall be susceptible at different times. The notion of the critical period has come of age and may be accepted as a legitimate concept in the study of behavior.

REFERENCES

1. Bickel, H., and Grueter, W. The dietary treatment of phenylketonuria—experiences during the past nine years. In P. W. Bowman and H. V. Mautner (Eds.), *Mental Retardation*, New York: Grune and Stratton, 1960.

2. Ford, F. R. *Diseases of the Nervous System in Infancy, Childhood and Adolescence* (4th Ed.) Springfield, Ill.: Thomas, 1960.

3. Friedmann, M., and Cohen, P. Agenesis of the corpus callosum as a possible sequel to maternal rubella during pregnancy. *Am. J. dis. child*, 73:178 thru ff., 1947.

4.   Holt, K. S. Growth disturbances. *Little club clinics in developmental medicine,* **4**:39-53, 1961.

5.   Labhart, A. *Klinik der inneren Sekretion.* Berlin: Springer-Verlag, 1957.

6.   Lenneberg, E. H. *Biological Foundations of Language.* New York, Wiley, 1967.

7.   Lyon, G. First signs and mode of onset of congenital hemiplegia. *Little club clinics in developmental medicine,* **4**:33-38, 1961.

8.   Norman, R. M. Hemiplegia due to birth injury. *Little club clinics in developmental medicine,* **4**:11-17, 1961.

9.   O'Connel, E. J., Feldt, R. H., and Stichter, G. B. Head circumference, mental retardation and growth failure. *Pediatrics,* **36**:62-66, 1965.

10. Paine, R. Disturbance of sensation in cerebral palsy. *Little club clinics in developmental medicine,* **2**:105-109, 1960.

11. Powell, G. F., Brasel, J. A., and Blizzard, R. M. Emotional deprivation and growth retardation simulating idiopathic hypopituitarism. I Clinical evaluation. *New Engl. J. Med.,* **276**:1271-1278, 1967.

12. Powell, G. F., Brasel, J. A., Raiti, S., and Blizzard, R. M. Emotional deprivation and growth retardation simulating idiopathic hypopituitarism. II Endocrinologic evaluation. *New Engl. J. Med.,* **276**:1279-1283, 1967.

13. Schaller, George B. *The Mountain Gorilla.* Chicago, University of Chicago Press, 1963.

14. Schultz, Adolph. Postembryonic age changes. In H. Hofer, A. H. Schultz, and D. Starck (Eds.), *Primatologia,* New York, S. Karger, 1956.

15. Smith, D. W., Blizzard, R. M., and Wilkins, L. The mental prognosis in hypothyroidism in infancy and childhood. A review of 128 cases. *Pediatrics,* **19**:1011 thru ff., 1957.

16. Stoch, M. B., and Smythe, P. M. Does undernutrition during infancy inhibit brain growth and subsequent intellectual development? *Arch. dis. child,* **38**:546-552, 1963.

17. Teuber, H. L., Battersby, W. S., and Bender, M. B. Effects of cerebral lesions on intellectual functioning in man. *Fed. proc. Am. soc. exp. biol.,* **11**:161 thru ff., 1952.

18. Tizard, P. Observations on the early manifestations of infantile hemiplegia. *Little club clinics in developmental medicine,* **4**:30-32, 1961.

19. Woods, G. Natural history of hemiplegia. *Little club clinics in developmental medicine,* **4**:26-29, 1960.

20. Zangwill, O. L. Intelligence in aphasia. In A. V. S. de Reuck and M. O'Connor (Eds.), *Disorders of Language,* Ciba Foundation Symp., Boston: Little, Brown, 1964.

NAME INDEX

171

SUBJECT INDEX

Abnormal environments, effects in man, 149
Activity, and hormones, 14
Amygdala, lesions of, 122, 123, 140
Animals, isolated, 6
  socially reared, 6
Anorexia, following frontal-caudate lesions, 132, 154
Apathy, 154
Aphasia, receptive, 165, 166
Athyroidism, 154, 156
Autoradiographic techniques, 54

Behavior, aggressive, 24, 27, 125
  circling, 125
  differences, in terms of siblings, litter-mates, 69
  "gape," 27
  general effect of frontal lobe damage, 99
  hippocampal lesions, effects on passive avoidance problem, 68
  open-field, 16, 17
  play, 26, 27
  runway problems, 66
  see also Sexual behavior
Bilirubin, 159, 160
Blindness, 61
Brain damage, and intelligence, 167
  reactions to in man, 161
  traumatic damage, 166
Brain lesion effects, compensable, 77
  noncompensable, 77
Brodmann areas, 9, 10, 11, 12, 165

Castration, neonatal, 24
Cat, ovariectomized, 10
  post-surgical development, 42
Caudate, 131
  frontal lesions, and, 132, 133, 134, 138, 140, 141
  nucleus, 88, 124, 129, 130, 133, 160
Cells, basophilic, of anterior pituitary, 8
  division, after brain damage, possibility of, 50
  of hippocampus, abnormal patterns of, 47, 48, 49

Cells *(Continued)*
  migration of, 162
  organization, abnormal, 61
Cerebellum, 160
Chickens, 9
  capons, 9
Circling behavior, 132, 141
Claustrum, 130
Convulsions, 99
Corpora lutea, formation of, 10
Cortex, areas, vicarious functioning of, 107
  auditory, 121
  combined frontal and posterior association area lesions, 135-142
  frontal association, 79-81, 84-89, 97, 99-103, 112-116, 122-124, 126-129, 131-134, 136, 138-140
  front parietal area, 166
  motor, 121
  occipital, 81
  orbitofrontal, 124, 130
  parietal, 162
  posterior association, 79, 80, 107, 122, 135-139
  premotor, 121
  somesthetic, 121
  temporal lobes, 81
  visual, 121
Cretinism, 153-158

Deafness and intelligence, 166
Dentate gyrus, 48
Development, critical periods, 168, 169
  genetic factors, 168
Developmental psychology, 1
Diencephalon, 163
Distractability, 132, 135
Dog, micturitional patterns of, 31, 32
Dwarfs, pituitary, 149

Emotional changes, following association cortex lesions, 138
Emotionality, 17
Erythroblastosis fetalis (RH incompatibility), 159

175